An Atlas of
PARKINSON'S DISEASE
AND RELATED DISORDERS

THE ENCYCLOPEDIA OF VISUAL MEDICINE SERIES

An Atlas of
PARKINSON'S DISEASE
AND RELATED DISORDERS

G. David Perkin, BA, FRCP

Regional Neurosciences Centre, Charing Cross Hospital
London, UK

Foreword by

Anthony E. Lang, MD, FRCPC

Director, The Toronto Hospital
Morton & Gloria Shulman Movement Disorders Centre
Toronto, Ontario, Canada

The Parthenon Publishing Group
International Publishers in Medicine, Science & Technology

NEW YORK   LONDON

Library of Congress Cataloging-in-Publication Data
Perkin, G. David (George David)
    An atlas of Parkinson's disease and related disorders / G. David Perkin ;
foreword by Anthony E. Lang.
        p.   cm. -- (The Encyclopedia of visual medicine series)
    Includes bibliographical references and index.
    ISBN 1-85070-943-2
    1. Extrapyramidal disorders--Atlases. 2. Parkinsonism--Atlases.
3. Movement disorders--Atlases. I. Title. II. Series.
    [DNLM: 1. Parkinson Disease--atlases. 2. Basal Ganglia Diseases--atlases.
3. Movement Disorders--atlases. WL 17 P447ac 1997]
RC376.5.P475 1997
616.8'33--dc21
DNLM/DLC
for Library of Congress   97-37265
                                                                        CIP

British Library Cataloguing in Publication Data
Perkin, G. David (George David)
    An atlas of Parkinson's disease and related disorders. -
    (The encyclopedia of visual medicine series)
    1. Parkinsonism
    I. Title
    616.8'33
    ISBN 1-85070-943-2

Published in the USA by
The Parthenon Publishing Group Inc.
One Blue Hill Plaza
PO Box 1564, Pearl River
New York 10965, USA

Published in the UK and Europe by
The Parthenon Publishing Group Limited
Casterton Hall, Carnforth
Lancs. LA6 2LA, UK

Copyright ©1998
Parthenon Publishing Group

Printed and bound in Spain
by T.G. Hostench, S.A.

Contents

Foreword

Since the widespread use of videotape, the neurological subspecialty of movement disorders has established a wide appeal and following, as evidenced by the avid atttendance of neurologists at 'unusual movement disorders' videotape sessions held at international meetings and the establishment of an international journal, *Movement Disorders*, which is accompanied by a videotape supplement.

In this era of multimedia, it is important that the illustrative power and specific advantages provided by still photography not be forgotten. There is a long and illustrious history of the depiction of disorders of movement and posture through the use of drawings and still photographs, as exemplified by the work of Charcot and his pupils at L'Hôpital de la Salpetrière in Paris in the late 1800s.

It is in this tradition that Dr David Perkin has compiled a modern series of still photos highlighting various aspects of Parkinson's disease and related motor disorders. This book provides a useful sample of clinical, investigative (CT, MRI and PET) and pathological images with a succinct descriptive text of the disorders featured. *An Atlas of Parkinson's Disease and Related Disorders* is an excellent introduction to this fascinating topic, and should serve as a stimulus to medical students and neurologists in training to pursue further studies in the field. This work will also serve as a useful adjunct to teaching videotapes of movement disorders which are capable of presenting the clinical features from a unique perspective, but are unable to demonstrate such aspects as imaging and pathology, which are so well represented in this atlas.

It is hoped that, stimulated by this book in combination with these other sources of information, a future generation of physicians will pursue studies designed to unlock the 'dark basements' of the brain (the basal ganglia) which contribute to these unusual and fascinating disorders of motor control.

Anthony E. Lang, MD, FRCPC
Toronto

Preface

In writing *An Atlas of Parkinson's Disease and Related Disorders*, I have been conscious of the need to find an appropriate match between the text and the illustrative material. The text is designed to provide a basic overview of the conditions discussed, inevitably concentrating on those areas which lend themselves best to photographic illustration. Some movement disorders, by their very nature, do not lend themselves to still photography whereas others, characterized by sustained postures, are ideally suited to the technique. Perhaps nowhere else in neurology is there such an opportunity to blend patient material, pathology and imagery in the discussion of the constituent conditions.

The development of brain-bank facilities such as the Parkinson's Disease United Kingdom Brain Bank has provided new insight into the spectrum of pathological entities underlying a particular clinical presentation while, at the same time, demonstrating that specific neuropathological entities may present with a considerable range of clinical features.

Accordingly, approximately one-third of the material in this atlas is pathological, incorporating both macroscopic and microscopic sections. A further quarter of the material is represented by imaging, principally magnetic resonance imaging (MRI) and positron emission tomography (PET) scanning. The area of movement disorders has been particularly fruitful for PET scanning, which promises, with the development of specific ligands for the various receptor sites, to further expand understanding of the pathophysiological mechanisms of the movement disorders.

It is expected that this atlas will provide a stimulating insight into the various aspects of the movement disorders for neurologists in training, but its approach to the subject should make it equally accessible for the medical student with an interest in neurological disorders.

It is a great pleasure to record the generosity of all the contributors who have provided me with material. I am particularly indebted to Dr Susan Daniel, who has been largely responsible for the superb pathological material in this atlas. I would also like to express a debt of gratitude to Dr M. Savoiardo who, not for the first time, has come to my rescue by providing state-of-the-art imaging material of many of the conditions discussed in the following pages.

G. David Perkin
London

Acknowledgements

I would like to thank the following publishers and authors, who have kindly allowed me to reproduce the following illustrations:

Figure 1, reproduced with permission of Harcourt–Brace and Dr C.G. Gerfen, modified from Figure 1 in Gerfen & Engber, Molecular neuroanatomic mechanisms of Parkinson's disease: A proposed therapeutic approach. *Neurol Clin* 1992;10:435–49

Figure 2, reproduced with permission of Lippincott–Raven and Dr C.G. Goetz, modified from a figure in Goetz *et al.*, Neurosurgical horizons in Parkinson's disease. *Neurology* 1993;43:1–7

Figure 19, reproduced with permission of Lippincott–Raven and Professor O. Lindvall, first published as Figure 5 in Lindvall *et al.*, Evidence for long-term survival and function of dopaminergic grafts in progressive Parkinson's disease. *Ann Neurol* 1994;35:172–80

Figures 24 and 25, reproduced with permission of Lippincott–Raven and Dr G. Fénelon, first published as Figures 1A and 2A in Fénelon *et al.*, Parkinsonism and dilatation of the perivascular spaces (état criblé) of the striatum: A clinical, magnetic resonance imaging, and pathological study. *Mov Disord* 1995;10:754–60

Figure 42, reproduced with permission of Lippincott–Raven and Dr S. Gilman, first published in Gilman *et al.*, Patterns of cerebral glucose metabolism detected with positron emission tomography differ in multiple system atrophy and olivopontocerebellar atrophy. *Ann Neurol* 1994; 36:166–75

Figure 50, reproduced with permission of Lippincott–Raven and Dr E.R.P. Brunt, first published as Figure 1b in Brunt *et al.*, Myoclonus in corticobasal degeneration. *Mov Disord* 1995;10: 132–42

Figure 55, reproduced with permission of Rapid Science and Dr J. Jankovic, first published as Figure 1 in Jankovic, Botulinum toxin in movement disorders. *Curr Opin Neurol* 1994;7:358–66

Figures 64 and 65, reproduced with permission of Lippincott–Raven and Dr A.E. Lang, first published as Figure 2 A and B in Jog & Lang, Chronic acquired hepatocerebral degeneration: Case reports and new insights. *Mov Disord* 1995;10:714–22

Figures 70 and 71, reproduced with permission of the American Roentgen Ray Society and Dr J.P. Comunale Jr, first published as Figure 1 B and C in Comunale *et al.*, Juvenile form of Huntington's disease: MR imaging appearance. *AJR* 1995; 165:414–5

Figure 72, reproduced with permission of Oxford University Press and Dr N. Turjanski, first published as Figure 2 in Turjanski *et al.*, Striatal D1 and D2 receptor binding in patients with Huntington's disease and other choreas: A PET study. *Brain* 1995;118:689–96

I am also indebted to the following colleagues, who have generously provided me with their unpublished material:

Figures 3–5, 26, 27, 32, 33, 36–39, 43, 45, 46, 61–63, 67 and 68, from Dr Susan E. Daniel, Senior Lecturer in Neuropathology and Head of Neuropathological Research, The Parkinson's Disease Society Brain Research Centre, Institute of Neurology, London, WC1N 1PJ

Figures 30, 35, 40, 41, 49 and 73, from Dr M. Savoiardo, Consultant Neuroradiologist, Department of Neuroradiology, Istituto Nazionale Neurologico "C. Besta", Milan, Italy

Figures 13, 74 and 75, from Dr P. Bain, Senior Lecturer in Clinical Neurology, The West London Neurosciences Centre, Charing Cross Hospital, London, W6 8RF

Figures 66 and 69, from Dr N. Wood, Senior Lecturer in Clinical Neurology, The Institute of Neurology, Queen Square, London, WC1N 3BG

Figures 21 and 22, from Dr D. Miller, Associate Professor of Neuropathology and Neurosurgery, NYU Medical Center, New York, and Professor M.H. Mark, The University of Medicine and Dentistry of New Jersey, New Jersey

Figures 20 and 34, from Dr D. Miller, Associate Professor of Neuropathology and Neurosurgery, NYU Medical Center, New York

Figure 6, from Dr W.R.G. Gibb, Consultant Neurologist, Institute of Psychiatry, London SE5 8AF

Finally, it is a pleasure to acknowledge my indebtedness to Dr Anthony Lang, both for his gracious introduction to my book as well as for his very helpful suggestions regarding the composition of the text and illustrations.

Section 1 A Review of Parkinson's Disease and Related Disorders

Anatomy

The neurons of the corpus striatum receive an excitatory input from the cerebral cortex and the thalamus. The major outputs project to the globus pallidus and the substantia nigra pars reticulata (SNr), and use gamma-aminobutyric acid (GABA) as a transmitter. Major efferent pathways from the globus pallidus interna and the SNr project to the thalamus. Feedback to the striatum is through the dopaminergic striatonigral pathway originating in the substantia nigra pars compacta (SNc; Figure 1).

These separate pathways use different neuropeptides and dopamine receptors. The direct pathway from the striatum to the globus pallidus interna (GPi) and SNr expresses substance P and dynor-phin, and uses D_1 dopamine receptors. The neurons projecting from the striatum to the external segment of the globus pallidus (GPe) express enkephalin and use D_2 receptors. (Some neurons express both receptors.) Depletion of dopamine in the striatum results in increased activity of this striatopallidal pathway and decreased activity in the direct pathway. These effects (the former leading to disinhibition of the subthalamic nucleus) lead to increased activity of the GABAergic neurons of the output nuclei of the basal ganglia. Increased inhibitory output from these nuclei may be responsible for the bradykinesia seen in patients with Parkinson's disease (Figure 2).

Parkinson's disease

Any discussion of the clinical characteristics of Parkinson's disease must take into account the inaccuracies of clinical diagnosis. In a successive series of 100 patients with a clinical diagnosis of Parkinson's disease, only 76 fulfilled the criteria for diagnosis at post-mortem examination (Table 1). Attempts to tighten the diagnostic criteria lead to increased specificity but with reduced sensitivity.

Neuropathology

Typically, there is loss of at least 50% of the melanin-containing nerve cells of the substantia nigra, the changes concentrating in the central part of the zona compacta (Figure 3). Accompanying these changes is depletion of tyrosine hydroxylase, the

Table I Pathological findings in 100 successive Parkinsonian patients

Idiopathic Parkinson's disease	76
Progressive supranuclear palsy	6
Multiple system atrophy	5
Alzheimer's disease	3
Alzheimer-type pathology with striatal involvement	3
Lacunar state	3
Nigral atrophy	2
Postencephalitic Parkinsonism	1
Normal (?essential tremor)	1

from Hughes *et al.*, 1992

rate-limiting enzyme in the biosynthetic pathway for catecholamines. (Figures 4 and 5). A characteristic, indeed inevitable, finding is the presence of Lewy bodies in some of the remaining nerve cells (Figure 6).

Together with Lewy body formation, degenerative changes occur at other sites, including the locus ceruleus, the dorsal motor nucleus of the vagus, the hypothalamus, the nucleus basalis of Meynert and the sympathetic ganglia. Cortical Lewy bodies are probably present in all patients with idiopathic Parkinson's disease, although not with the frequency that would permit a diagnosis of cortical Lewy body disease (*vide infra*).

In Parkinsonian patients with cortical dementia, the pathological changes are either those of cortical Lewy body disease, or those associated with Alzheimer's disease, including senile plaques, neurofibrillary tangles, granulovacuolar degeneration, and nerve cell loss in the neocortex and hippocampus.

Epidemiology

The prevalence of Parkinson's disease has been reported to lie between 30 and 300 / 100 000, producing approximately 60 to 80 000 cases in the United Kingdom. Prevalence increases with age and the disease is slightly more common in men (Figure 7). Cigarette-smoking provides some protective effect, whereas the risk is increased in those with a history of herbicide or pesticide exposure.

Clinical features

Typically, the condition produces bradykinesia, tremor, rigidity and impairment of postural reflexes. An asymmetrical onset is characteristic.

Bradykinesia

Paucity of movement can affect any activity and is best measured by assessing aspects of daily living. The problem tends to involve one upper limb initially, leading to difficulty with fine tasks, such as manipulating a knife or fork, dressing or shaving. The patient's handwriting typically becomes reduced in size if the dominant hand is affected (Figure 8). Associates are likely to comment on a reduction of arm swing when walking. Facial immobility is evident, with a lack of animation and immediate emotional response (Figure 9). The posture is stooped, and becomes more so as the condition progresses (Figures 10 and 11). Walking becomes slowed, with a tendency to reduce stride length and an increased number of steps being taken when turning. The problem can be assessed by asking the patient to repetitively tap with the hand or foot, or to mimic a polishing motion with the hand, or to rhythmically clench and unclench the fingers (Figure 12). Even if the amplitude of such movements is initially retained, it soon diminishes and may even cease.

Rigidity

The rigidity associated with Parkinson's disease is also often asymmetrical at onset. It tends to be diffusely distributed throughout the limb although, initially, it may be more confined. It persists throughout the range of motion of any affected joint. A characteristic judder (cogwheeling) occurs at a frequency similar to that of the postural tremor seen in Parkinson's disease rather than at the rate of the resting tremor. If the rigidity is equivocal, it can be activated by contracting the contralateral limb.

Tremor

The classical Parkinsonian tremor occurs at rest, at a frequency of around 3–4 Hz (Figure 13). The tremor briefly inhibits during a skilled activity. A faster, postural tremor of around 6–8 Hz is sometimes evident initially at a time when the rest tremor is absent. The rest tremor most commonly involves the upper limb, producing either flexion / extension movements or pronation / supination, or a combination of these.

Postural reflexes

In addition to abnormalities of posture, the patient has difficulty maintaining posture when suddenly pushed forwards or backwards. Other features of Parkinson's disease include dementia (perhaps in around 15–20% of patients), autonomic dysfunction (principally in the form of urinary urgency and occasional incontinence) and a variety of eye signs, including broken pursuit movements and some limitation of upward gaze and convergence. A positive glabellar tap (producing repetitive blinking during tapping over the glabella) occurs in the majority, but is also seen in Alzheimer's disease (Figure 14).

Imaging

Although imaging techniques, particularly positron emission tomography (PET) scanning, are not relevant to the diagnosis of most patients with Parkinson's disease, they do provide insight into the pathophysiology of the disease and can assume clinical relevance where the clinical presentation is atypical. PET scans using 6-[^{18}F]-fluorodopa show reduced uptake of the isotope, particularly in the putamen and mainly contralateral to the clinically more affected side (Figure 15).

Drug intervention

There are potentially several stages during the synthesis, release and metabolism of dopamine within the central nervous system at which intervention, by enhancing dopamine levels, may influence the clinical manifestations of Parkinson's disease.

Dopa is converted to dopamine within the dopaminergic neuron by the action of L-aromatic-amino-acid decarboxylase (dopa decarboxylase). The dopamine is then transported into storage vesicles before being released, through depolarization and entry of calcium ions, to act on the postsynaptic dopamine-receptor site. Some of the dopamine is taken up again in the dopaminergic neuron while another part is converted, within glial cells, to 3-methoxytyramine by the action of catechol *O*-methyltransferase (COMT). The 3-methoxytyramine is then metabolized by monoamine oxidase-B to homovanillic acid (HVA). Some of the dopamine that is taken up again into the neuron is transported back into storage vesicles, whereas the remainder is metabolized by monoamine oxidase-B to 3,4-dihydroxyphenyl-acetic acid (DOPAC). Dopaminergic activity can

therefore be enhanced by providing more precursor (dopa; Figure 16), stimulating dopamine release (amantadine), using an agonist to act on the dopamine-receptor site (bromocriptine, lysuride, pergolide, ropinirole and cabergoline) or inhibiting dopamine breaknown through inhibition of either monoamine oxidase (selegiline) or of COMT (tolcapone).

Dopa, combined with a dopa-decarboxylase inhibitor, remains the cornerstone of treatment. The use of subcutaneous apomorphine as a diagnostic test for idiopathic Parkinson's disease has been advocated, but both false-positive and false-negative results occur. There is no consensus as to whether agonist therapy should be introduced earlier or later. After 5–10 years, major therapeutic problems arise, with loss of efficacy, fluctuations in response and the emergence of increasingly uncontrollable dyskinesias or dystonic posturing (Figures 17 and 18). These problems have stimulated consideration of other therapeutic approaches, including thalamic (Figure 19) and pallidal surgery, and transplantation of dopaminergic grafts. Such grafts, derived from human embryonic mesencephalic tissue, have been shown to have a functional effect for at least 3 years after transplantation, as substantiated by evidence of enhanced putaminal fluorodopa uptake over the same period (Figure 20).

Parkinsonian syndromes

A vast number of disorders can produce a clinical picture which closely resembles Parkinson's disease (Table 2).

Postencephalitic Parkinsonism

Cases of postencephalitic Parkinsonism still occur sporadically. Besides the Parkinsonism, clinical features include oculogyric crises, behavioral disorders, pyramidal tract signs and various movement abnormalities. Depigmentation of the substantia nigra is evident, along with the presence of neurofibrillary tangles. Although inflammatory cells are conspicuous in the acute stage, they may still be present years later.

Drug-induced Parkinsonism

Any drug affecting the synthesis, storage or release of dopamine, or interfering with dopamine receptor sites, is capable of causing an akinetic rigid syndrome which may closely resemble idiopathic Parkinson's disease. The most well-recognized drugs in this category are the phenothiazines but, in addition, a calcium-blocking vasodilator such as flunarizine or the antihistamine cinnarizine can induce Parkinsonism, possibly through a presynaptic effect on dopaminergic and serotonergic neurons.

The condition tends to be symmetrical and to lack tremor. If a tremor is present, it tends to be postural and of a higher frequency than the classical resting tremor of idiopathic Parkinson's disease. Most cases are evident within 3 months of starting therapy.

The problem is more likely to affect the elderly and women, and may take several months to subside after drug withdrawal. If the symptoms are disabling and the drug therapy is still required, either amantadine or an anticholinergic agent has been suggested as appropriate treatment.

Table 2 Disorders with clinical presentations similar to Parkinson's disease

Symptomatic Parkinsonism
  Postencephalitic
  Drug-induced
  Toxic
  Traumatic
  Arteriosclerotic
  Normal-pressure hydrocephalus
  Striatonigral degeneration

Parkinsonism in other degenerative disorders
  Multiple system atrophy
  Progressive supranuclear palsy
  Corticobasal degeneration
  Diffuse Lewy body disease

Arteriosclerotic Parkinsonism

Parkinsonian features are sometimes part of the clinical spectrum associated with diffuse cerebrovascular disease. In the original description, certain clinical features were held to distinguish arteriosclerotic Parkinsonism from idiopathic Parkinson's disease, including the lack of tremor, a predominance of gait involvement over upper limb disorder and the presence of signs in other systems, for example, bilateral extensor plantar responses. In such patients, particularly those with a history of hypertension or stroke-like events, the possibility of a Binswanger-type encephalopathy as the underlying mechanism is considerable (Figure 21).

Microscopy reveals sharply defined zones of myelin loss (Figure 22), with or without coexistent areas of lacunar infarction (Figure 23). Either pathology is usually demonstrable with appropriate imaging (Figure 24).

Some patients with a Parkinsonian state due to vascular disease have rest tremor whereas others show dopa responsiveness. Whether expanded perivascular spaces alone (*état criblé*) within the striatum can be responsible for a Parkinsonian state is still under debate. If this is the case, the clinical picture is then atypical for idiopathic Parkinson's disease with the presence of predominant axial involvement (Figures 25 and 26).

Cortical Lewy body disease

The prevalence of a cortical-type dementia in Parkinson's disease has long been debated. Most of the recent surveys give a figure between 15–20% of the population.

Risk factors for dementia in Parkinsonian patients include age and duration of the disease. In some Parkinsonian patients with dementia, post-mortem examination establishes the presence of neurofibrillary tangles, granulovacuolar degeneration, and nerve cell loss in the hippocampus and neocortex of a nature consistent with a diagnosis of Alzheimer's disease. In other patients, the major cortical pathology is the presence of Lewy bodies (Figure 27).

Occasional cortical Lewy bodies can probably be found in all Parkinsonian patients but, where the bodies are profuse and widely scattered in the neocortex, a differing clinical pattern emerges, described as diffuse Lewy body disease or Lewy body dementia. Additional pathological features include spongiform degeneration and ubiquitous immunoreactive neurites in parts of the hippocampus. To further complicate the classification of this entity, perhaps as many as half the patients with cortical Lewy body disease have concomitant Alzheimer pathology.

In patients with Lewy body dementia, the dementia may precede, coincide with or follow the extrapyramidal features. Early onset of paranoid ideation accompanied by visual hallucinations in a Parkinsonian patient is suggestive of the diagnosis. Falls are commonplace. The Parkinsonian features may or may not be responsive to dopa therapy.

Related disorders

Progressive supranuclear palsy (Steele–Richardson–Olszewski syndrome)

For many, or perhaps even all, extrapyramidal syndromes, a classical picture is described which is anticipated to predict a particular pathological entity at post-mortem examination. As knowledge of the disease grows, however, it soon becomes apparent that the same disease process – as defined pathologically – has a much broader clinical spectrum than was appreciated in the original description. The converse also applies: patients with a classical clinical syndrome may prove to have other pathological entities.

Nowhere are these discrepancies more evident than in cases of progressive supranuclear palsy (PSP). One of the problems in establishing clinicopathological correlations in PSP is the lack of consensus as to the pathological criteria for the diagnosis. Certain features, however, are predictable. The substantia nigra shows severe pigment depletion as does the locus ceruleus. Neuronal loss is found in the substantia nigra, subthalamus and globus pallidus. Neurofibrillary tangles can be identified in the cerebral cortex, caudate, putamen, globus pallidus, subthalamus and brain stem (Figure 28). Accompanying the neurofibrillary tangles are neuropil threads (silver- and tau-positive). Typically, changes are found in the regions associated with vertical gaze, including the rostral interstitial nucleus of the medial longitudinal fasciculus and the interstitial nucleus of Cajal.

A disturbance of gait is common and many patients are liable to falls. The body tends to remain extended rather than taking on the stooped posture of Parkinson's disease. Pseudobulbar features are prominent, with dysphagia, dysarthria and emotional incontinence. The supranuclear palsy first affects down gaze, and particularly downward saccades (Figure 29). Some patients complain of blurred vision or frank diplopia. Later, vertical, then horizontal, saccades become compromised followed by impairment of pursuit movement. Reflex eye movements, elicited by the doll's-head maneuver, are spared initially (Figure 30), but are later lost so that a total ophthalmoplegia becomes evident. In well-documented cases, despite the appropriate pathological changes found post-mortem, the patient may have had no disturbances of eye movements in life. Limb rigidity is less prominent than axial rigidity. Bradykinesia is present to a varying degree with some patients presenting as a pure akinetic syndrome. Tremor occurs in around 12–16% of cases. A subcortical, rather than cortical, dementia is characteristic.

In most cases, dopa therapy is ineffective and almost never influences the ophthalmoplegia.

Imaging changes include both generalized and selective brain stem atrophy (Figure 31). Single photon emission computed tomography (SPECT) can demonstrate impairment of frontal perfusion with an intact cortical rim. PET scanning shows decreased metabolic activity in the frontal cortex, caudate and putamen together with evidence of abnormal D_2-receptor function (Figure 32).

Striatonigral degeneration

This condition is frequently confused with Parkinson's disease in life. At post-mortem, there is atrophy and discoloration of the putamina (Figure 33) accompanied, in almost half the cases, with atrophy of the caudate nuclei. The changes in the putamen begin dorsally in the posterior two-thirds, then spread ventrally and anteriorly. On microscopy, the putamen shows intracellular pigmentation, gliosis and loss of myelinated fibers (Figure 34). Neuronal depletion, gliosis and loss of myelinated fibers are seen in the globus pallidus whereas both the substantia nigra and locus ceruleus show pallor with microscopic evidence of neuronal loss and gliosis (Figure 35). Lewy bodies are seldom found. In some cases, even without clinical features in life, there is involvement of the olivopontocerebellar system.

Striatonigral degeneration has considerable clinical overlap with Parkinson's disease, but sufficient differences to suggest the diagnosis in life. Rest tremor in the early stages of the disease is distinctly uncommon, although it appears in half of the cases during the later stages of the disease. The condition is equally likely as Parkinson's disease to be asymmetrical at onset. Falls early in the course of the disease are a recognized feature. Some patients show a response to dopa. Other features which should suggest the diagnosis

include severe dysphonia and dysphagia, and the development of autonomic symptoms or cerebellar signs, indicating the development of multiple system atrophy (*vide infra*).

On T_2-weighted magnetic resonance imaging (MRI), low signal intensity is seen in the putamen, sometimes bordered by a thin rim of hyperintensity (Figure 36). PET scanning can demonstrate reduced striatal and frontal lobe metabolism.

Multiple system atrophy

Autonomic features may accompany a Parkinsonian syndrome without evidence of other system involvement. In such patients, the autonomic failure is due to intermediolateral column degeneration in the spinal cord whereas the Parkinsonian syndrome reflects the classical features of idiopathic Parkinson's disease, including typical changes in the substantia nigra and locus ceruleus, with Lewy body formation. In other patients, described as having multiple system atrophy, the autonomic failure is due to the same pathological process in the spinal cord, but the other clinical features represent a combination, in varying degrees, of striatonigral degeneration and olivopontocerebellar atrophy (OPCA).

In OPCA, there is macroscopic evidence of atrophy of the pons, middle cerebellar peduncle, parts of the cerebellum and the olives (Figure 37). Microscopically the pontine tegmentum is virtually spared, but there is pallor of the transverse fibers in the basis pontis together with neuronal loss (Figure 38). Depletion of both granules and Purkinje cells is seen in the cerebellum. Where the latter has occurred, empty 'baskets' with hypertrophied fibers are seen associated with the formation of axon 'torpedoes' in the molecular layer (Figure 39). Oligodendroglial cytoplasmic inclusions are probably seen in all cases of multiple system atrophy and in all sporadic cases of

OPCA, but only rarely in familial cases of OPCA (Figure 40).

Clinical criteria have been suggested for the diagnosis of multiple system atrophy (Table 3). Diagnostic problems arise as the result of some patients who present with Parkinsonism, others who have a cerebellar syndrome, and a third group who manifest autonomic failure, without clear evidence in all three instances of other system involvement. Sporadic cases are not seen in those under 30 years of age. Dementia is not a feature of multiple system atrophy, nor is there an ophthalmoplegia (although this is recorded in both sporadic and familial forms of OPCA). Although poor or absent dopa responsiveness is the norm, some cases – confirmed at post-mortem examination – may

show a response comparable to that seen in idiopathic Parkinson's disease.

Multiple system atrophy usually presents in the sixth decade of life. The median survival is of the order of 7–8 years. Men are slightly more often affected than women. The most common combination of clinical features is autonomic impairment with Parkinsonism. Autonomic symptoms include postural hypotension, urinary urgency with incontinence and erectile failure in male patients. Fecal incontinence is uncommon and syncopal attacks are a feature in only a minority of cases. Speech impairment is almost inevitable, with a combination of dysarthria and dysphonia producing a variety of speech disorders. Overall, cerebellar signs are recorded in nearly half the cases, and pyramidal

Table 3 Multiple system atrophy: Proposed clinical diagnostic criteria

	Striatonigral type (predominantly Parkinsonism)	Olivopontocerebellar type (predominantly cerebellar)
Definite	Post-mortem confirmation	Post-mortem confirmation
Probable	Sporadic adult-onset	Sporadic adult-onset
	Non- or poorly levodopa-responsive Parkinsonism	Cerebellar syndrome (with or without Parkinsonism or pyramidal signs)
	PLUS	PLUS
	severe symptomatic autonomic failure	severe symptomatic autonomic failure
	OR	OR
	cerebellar signs	pathological sphincter electromyogram
	OR	
	pyramidal signs	
	OR	
	pathological sphincter electromyogram	
Possible	Sporadic, adult-onset, non- or poorly levodopa-responsive Parkinsonism	Sporadic adult-onset cerebellar syndrome with Parkinsonism

Adult-onset; ≥ 30 years of age;
Sporadic; no multiple system atrophy in first- or second-degree relatives;
Autonomic failure; postural syncope and / or urinary incontinence or retention not due to other causes;
Levodopa-responsive; moderate or good levodopa-response accepted if waning and multiple atypical features present;
Parkinsonism; no dementia, areflexia or supranuclear down-gaze palsy

signs in almost two-thirds. Both bradykinesia and rigidity are likely, but a classical resting tremor is unusual. Even when the condition has presented in a pure cerebellar, Parkinsonian or autonomic format, it is never the case that that picture remains unaltered until death, except in the small percentage of cases with isolated Parkinsonism.

The good response to dopa, seen in a minority of cases, is seldom sustained. In such cases, substitution of a dopaminergic agonist is usually unhelpful. Drug-induced movements in these patients usually takes the form of dystonia rather than chorea. Certain other clinical features are suggestive of the disease and are notoriously difficult to manage. These include postural instability with falls, excessive snoring associated with vocal cord abductor palsy and anterocollis.

Imaging

Magnetic resonance imaging

MRI identifies sites of maximum atrophy in the brain stem and cerebellum. The middle cerebellar peduncle shows the most marked reduction in size, but other affected structures include the cerebellar vermis, the cerebellar hemispheres, the pons and the lower brain stem (Figure 41). Signal hyperintensities can be identified within the pons and middle cerebellar peduncles (Figure 42). Additional MRI findings include putaminal hypointensities. The relative distribution of the changes seen on MRI correlates, to a limited degree, with the clinical characteristics.

SPECT / PET

With the use of ^{123}I-iodobenzamide (IBZM)–SPECT, dopamine D_2 receptors can be imaged and shown to be significantly depleted in the striatum in patients with multiple system atrophy. PET using [^{18}F]-fluorodeoxyglucose has been used to measure local cerebral metabolic rates for glucose in both multiple system atrophy, and sporadic and familial forms of OPCA. In the former two, reduced metabolic activity, albeit to differing degrees, is found in the brain stem, cerebellum, putamen, thalamus and cerebral cortex. In familial OPCA, changes are confined to the brain stem and cerebellum (Figure 43).

Corticobasal degeneration

This disorder bears some superficial resemblance to PSP, but has distinctive clinical and pathological features which distinguish it. The gross pathological findings include a marked asymmetrical fronto-parietal atrophy with relative sparing of the temporal cortex (Figure 44). Both gray and white matter show gliosis and cell loss. Subcortical nuclei are also affected, with the most prominent changes being found in the substantia nigra. Other affected areas include the lateral thalamic nuclei, globus pallidus, subthalamic nuclei, locus ceruleus and red nucleus. A characteristic, but non-specific, finding is the presence of swollen achromatic neurons (balloon cells) in the affected cortical areas (Figures 45 and 46). A number of inclusion bodies have been found: those with a weakly basophilic body, called the corticobasal inclusion body; and small, more basophilic, bodies, which may represent a variant of the former rather than a distinct entity (Figure 47).

Typically, the condition begins insidiously and asymmetrically with a variety of motor deficits, including dystonia (Figure 48), an akinetic–rigid syndrome or the alien limb phenomenon. The affected upper limb takes on characteristic abnormal postures, particularly when the patient's attention is diverted or their eyes are closed. At times, the hand carries out relatively complex tasks when the patient is concentrating on other activities. In addition, the patient often shows features of an ideomotor or ideational apraxia (Figure 49). Other limb abnormalities include focal reflex myoclonus, other involuntary movements and grasp reflexes. A supranuclear eye-movement disorder similar to that seen in PSP may be present, or an apraxia of eye movement or eyelid opening. Postural instability is common, whereas falls and cortical sensory loss are found in around three-quarters of patients.

Computed tomography (CT) or MRI may demonstrate asymmetrical cortical atrophy (Figure 50). [^{18}F]-Fluorodopa–PET scanning shows striatal and cortical dopamine depletion. [^{18}F]-Fluorodeoxy-glucose–PET scanning demonstrates regional reduction in glucose metabolism (Figure 51). A comparison has been made between corticobasal degeneration and Pick's disease but, in most cases, there are sufficient clinical and pathological differences to establish the conditions as separate entities.

Dystonia

Torsion dystonia is a condition in which sustained muscle contraction leads to altered postures of the limb and trunk. The condition may be associated with other movement disorders, and is classified into a primary (idiopathic) form and various secondary (symptomatic) forms.

Idiopathic torsion dystonia may occur sporadically or in a genetically determined form, when it usually demonstrates autosomal-dominant transmission. The hereditary forms tend to present in children typically with involvement of one leg before progressing to the other limbs and the trunk. Dystonias can also be classified according to their distribution (Table 4).

Idiopathic dystonia usually starts in one leg, less commonly in the arm and least often in the trunk, particularly in cases presenting in the first decade of life. With a late presentation, initial involvement of the arm is more likely. With time, the condition spreads and accentuates.

Table 4 Classification of dystonia according to distribution

A. Generalized dystonia

B. Multifocal dystonia: affects two or more non-contiguous parts

C. Hemidystonia: Involvement of one arm and the ipsilateral leg

D. Segmental dystonia: either cranial (two or more parts of cranial and neck musculature), axial (neck and trunk), brachial (arm and axial or both arms ± neck ± trunk) or crural (one leg and trunk or both legs ± trunk)

E. Focal dystonia: affecting a single site such as eyelids (blepharospasm), mouth (oromandibular dystonia), larynx (spastic dysphonia), neck (torticollis) or arm (writer's cramp)

Fahn, Marsden & Calne, 1987

Typically, the foot tends to invert and plantar flex while involvement of the trunk produces a variety of abnormal body postures (Figures 52 and 53). Muscle tone is normal apart from the presence of active muscle contraction. Other clinical abnormalities are absent. No clear pathological substrate for idiopathic torsion dystonia has been found. Treatment for the condition is often disappointing, although anticholinergic therapy, in large doses, is sometimes beneficial. An occasional response is seen to dopaminergic agonists and antagonists, and benzodiazepines.

Focal dystonia

A variety of focal dystonias has been described. These tend to present in adult life and principally affect the muscles of the arm or neck, or those innervated by the cranial nerves. As with idiopathic torsion dystonia, focal pathological abnormalities have not been demonstrated *post mortem*.

Blepharospasm

This involves an increased blinking frequency which may culminate in the eyes becoming almost permanently closed (Figure 54). Sometimes a light touch to the eyelid may relieve the spasm, as may various diversionary physical actions on the part of the patient.

Oromandibular dystonia

This describes an abnormal movement of the jaw, mouth and tongue associated with dysphagia and dysarthria. The symptoms are typically triggered by attempts to speak or eat. Trauma to the tongue and buccal mucosa is a common occurrence.

Spasmodic dysphonia

Dystonia of the laryngeal muscles produces an abnormal voice pattern. Adduction of the vocal cords is seen more often than is abduction, and imparts a strained and harsh quality to the speech.

Spasmodic torticollis

Abnormal neck postures result from contraction of the sternocleidomastoid, splenius capitis, or both. There may be predominant rotation, or lateral flexion or extension. The condition may resolve, only to return later (Figure 55). A tremulous movement is often superimposed on a more sustained posture. Neck discomfort is common, and some patients develop degenerative disease of the cervical spine.

Writer's cramp

This is one of a number of occupational cramps in which dystonic posturing, frequently of a painful nature, develops in patients who use their hands habitually in performing a skilled task. Other activities associated with this condition include typing, playing the violin and cutting hair. The movements typically are generated only when a specific task is attempted. Other skilled activities of the hand are spared. Typically, excessive force is used, and the pen is held in an abnormal posture. The movement is often accompanied by inappropriate movement and posturing of the proximal arm muscles. Occasionally, the problem remits. Eventually, some patients learn to write with the other hand, although at the risk of then developing the problem in that hand as well.

Treatment

Treatment of the focal dystonias has been largely ineffective in the past, although certain dystonias (particularly blepharospasm and spasmodic torticollis) have shown a gratifying response to injections of botulinum toxin. There are several immunologically distinct forms of the toxin, of which type A is the most widely researched. Type A inhibits acetylcholine release from the presynaptic neuromuscular terminal by clearing synaptosomal-associated protein (SNAP-25; Figure 56). The consequent chemodenervation produces muscle paralysis and atrophy. Nerve sprouting and reinnervation occur over the following 2–4 months.

Secondary (symptomatic) dystonia

A vast array of conditions has been described as potential causes of secondary or symptomatic dystonia. These perhaps account for one-third of all cases. Although some patients present with pure dystonia, the majority have additional neurological abnormalities.

Certain characteristics point to the symptomatic forms of dystonia. Hemidystonia usually implies a structural lesion in the contralateral putamen or its connections. Perinatal hypoxia can lead to a number of movement disorders, including chorea, athetosis and dystonic posturing (Figures 57 and 58). In cases with a global failure of cerebral perfusion, pathological consequences include border-zone infarction together with ischemic changes in the putamen, thalamus and cerebellum. A more focal cerebral insult in the perinatal period may also be associated with focal dystonia and corresponding imaging abnormalities (Figures 59 and 60). Adult-onset ischemia is equally capable of producing a hemidystonic phenomenon that often appears following resolution of an initial hemiparesis (Figure 61).

Aspects of the clinical course also help to differentiate between the idiopathic and symptomatic forms of dystonia. Idiopathic forms tend to develop insidiously, are more or less progressive and only eventually lead to sustained dystonic postures. Symptomatic dystonias tend to develop more abruptly with sustained postures at an earlier age.

Wilson's disease

Wilson's disease is inherited as an autosomal-recessive trait. The prevalence of the condition is estimated to be 30 / 1 000 000 with the carrier state estimated to be 1% of the population. The disease is associated with a deficiency of serum ceruloplasmin. Impaired hepatic excretion of copper into bile leads to an abnormal accumulation of copper, initially in the liver and later in other organs. In some patients, the changes in the liver are non-specific in the form of a toxic hepatitis whereas, in others, a macro- and micronodular cirrhosis evolves, sometimes with no previous clinical evidence of liver disease.

Changes found in the brain include atrophy, softening and contraction of the basal ganglia, especially in the putamen. Changes are also found in cortical white matter, the cerebellar folia and the pons. Microscopically the putamen is atrophied and rarefied (Figure 62). The white matter shows spongy degeneration with loss of myelin fibers. Accumulation of type 1 and type 2 astrocytes (Figure 63) and Opalski cells is seen (Figure 64). The latter are of unknown origin. There is a surprisingly poor correlation between the degree of hepatic and cerebral damage and the clinical condition of the patient.

Neurological manifestations of the disease, which may be the presenting feature in nearly half the cases, appear from the second decade of age onwards, but rarely after the age of 40 years. The major declaration of the disease is in the form of involuntary movements coupled with prominent involvement of the facial and bulbar muscles. Abnormal movements principally consist of various forms of dystonic posturing. Chorea or choreoathetosis is uncommon. Dysarthria, which may partly be due to dystonia of the face and bulbar muscles, is prominent. Dysphagia is present and is accompanied by incessant drooling of saliva. A particular facial expression is described with retraction of the upper lip (*risus sardonicus*). On occasions, a more Parkinsonian picture emerges, with rigidity and tremor. The tremor is sometimes resting, at other times postural and, occasionally, of the so-called wing-beating type, describing a large-amplitude, violent, upper-limb tremor capable of causing trauma to the patient's own body. Cerebellar findings have also been identified, including limb and gait ataxias. A variety of eye-movement disorders has been described, but seldom proves to be symptomatic. Deposition of copper in Descemet's membrane of the cornea is probably inevitable in patients with neurological manifestations of Wilson's disease, but may require slit-lamp microscopy for identification.

Psychiatric manifestations are virtually ubiquitous, and may antedate other features of the disease. A profound psychotic state that is indistinguishable from schizophrenia is recognized, as are depressive states and severe behavioral disorders. Other organs that may be affected include the skin, the kidney and the skeleton.

The diagnosis can be confidently made if Kayser–Fleischer rings are identified. The vast majority of patients have a serum ceruloplasmin concentration <20 mg/dl. Urinary copper levels are usually high. Measurement of serum copper is unhelpful. On occasions, a liver biopsy with estimation of copper content is needed to establish the diagnosis.

Imaging is of value in demonstrating the particular changes occurring in the brain. CT can demonstrate ventricular dilatation and cortical atrophy as well as hypodensities in the basal ganglia. MRI is more sensitive in detecting both lesions within the basal ganglia and in the thalamus.

A chronic non-familial form of hepatic cerebral degeneration has been described. The clinical features are similar to those of Wilson's disease, but there are no Kayser–Fleischer rings, and no evidence of abnormal copper accumulation. The clinical features are variable and include an encephalopathic syndrome, various movement disorders and a myelopathy. The underlying hepatic disease may be silent. The condition is likely to coexist with episodes of acute hepatic encephalopathy, but its severity does not correlate with the frequency of such episodes. Indeed, in some cases, episodes of hepatic encephalopathy have not been reported. The initial presentation may be with either the hepatic or neurological features. As regards the movement disorder, dystonia is uncommon whereas chorea, and postural and action tremors, are often prominent. A variety of hepatic diseases appear capable of triggering acquired hepatocerebral degeneration, including chronic active hepatitis, primary biliary cirrhosis and other forms of intra- or extrahepatic portal–systemic shunt.

Both cerebral and cerebellar cortical atrophy can be demonstrated by CT scanning. MRI changes include hyperintense signals on T_1-weighted images in the globus pallidus, putamen and mesencephalon in the region of the substantia nigra (Figures 65 and 66).

The etiology of the brain lesions has not yet been established, although abnormal accumulation of manganese has been proposed as a possible factor. Some of the movement disorders may respond to dopa treatment.

Huntington's disease

The reported prevalence rates for this disease from the UK and USA have been 5–9 / 100 000. Although the disease most often appears in subjects in their late 30s and early 40s, onset in adolescence and over the age of 50 years is well recognized. A preponderance of juvenile-onset cases show male transmission. The Huntington gene has been localized to the short arm of chromosome 4. The gene displays an expanded and unstable trinucleotide repetition (37–86 repeat units in one series) compared with 11–34 copies in the normal chromosome. The age of onset of the disease is inversely correlated with the repeat length (Figure 67).

In terms of pathology, there is severe neuronal loss in the caudate and putamen and, to a lesser extent, in the globus pallidus and cerebral cortex. Macroscopically the brain is shrunken with widening of the cortical sulci and dilatation of the lateral ventricles (Figure 68). On microscopy, there is a marked depletion of striatal neurons which disproportionally affects small cells. Glial cell loss is less intense (Figure 69). The changes in the cortex are less substantial and are predominant in the third and fifth layers. A number of neurotransmitter systems is affected with particular depletion of GABA and acetylcholine.

Characteristic clinical features of the condition include chorea with intellectual decline and behavioral disorders. The onset is insidious. The chorea is often initially very subtle and may present in the limbs, axial muscles or muscles innervated by the cranial nerves. With time, dysarthria and dysphagia emerge together with an alteration of gait. Various eye movement changes are described, including abnormalities of pursuit and saccades. Intellectual changes affect the ability to plan and carry out sequential processes coupled with defects of memory and the ability to acquire new information. Behavioral abnormalities include lability, withdrawal and substantial changes in personality.

Juvenile cases (defined as onset before the age of 20 years) account for approximately 5% of cases and usually inherit the disease from affected fathers. In these cases, an akinetic–rigid syndrome is more likely than the classical presentation. At the other end of the age spectrum, Huntington's disease may also present atypically. Families are described in whom the disease usually presents after the age of 50 years and then in the form of chorea, with little or no evidence of dementia. Typically, these patients survive for much longer than classical cases. Furthermore, imaging fails to reveal evidence of disproportionate caudate or putaminal atrophy.

Imaging

CT reveals evidence of cortical and basal ganglia atrophy. A measure of caudate nuclear size (the bicaudate diameter) shows significant differences compared with a control population (Figure 70). The caudate and putaminal atrophy are better defined by MRI. In the classical form of the disease, abnormal signals from these nuclei are unusual. In the akinetic–rigid form, however, T_2-weighted images demonstrate increased signal intensity in both the caudate and the putamen (Figures 71 and 72). SPECT can demonstrate reduced striatal blood flow compared with controls. Post-mortem studies have established a reduction of both D_1 and D_2 receptors in the putamen. The radioactive tracer ^{11}C-raclopride is a selective reversible D_2-receptor antagonist whereas ^{11}C-SCH 23390 is a selective D_1-receptor antagonist. Using these tracers, Huntington's disease patients can be shown to have significant reductions in striatal D_1 and D_2 receptor density. The abnormalities apply both to the choreic and akinetic–rigid forms of the disease, but are greater in the latter group (Figure 73).

The condition is untreatable, although the movement disorder can be controlled, to some extent, by dopaminergic blockade. Isolation of the responsible gene has allowed accurate genetic counseling.

Hallervorden–Spatz disease

This rare disorder is usually familial with an autosomal-recessive inheritance. Onset is within the first two decades of life with disturbances of speech and gait. Extrapyramidal features predominate on examination, but with the addition of spasticity. Iron accumulates particularly in the substantia nigra and globus pallidus. MRI findings are characteristic, with diffuse low signal intensity on T_2-weighted images in the globus pallidus, accompanied by an anteromedial area of high signal intensity (eye-of-the-tiger sign; Figure 74).

Sydenham's chorea

This disease is one of the recognized manifestations of acute rheumatic fever. The chorea is accompanied by dystonia and often psychological symptoms, of which emotional lability is the most prominent. The psychological manifestations usually antedate the chorea. The condition usually presents at around 8–9 years of age and lasts for an average of 6 months. In some cases, the chorea is confined to one side of the body. Most children with

Sydenham's chorea have other manifestations of rheumatic fever, usually either arteritis or carditis. Chorea is estimated to occur in around 10–20% of patients with acute rheumatic fever. The condition is explicable on the basis of an antibody, triggered by group A beta-hemolytic streptococcal infection, which crossreacts with an unidentified antigen on neurons within the basal ganglia. The severity of the chorea can be correlated with the presence and titer of the antibody. Plasmapheresis or immunoglobulin therapy probably shortens the duration, and lessens the severity, of the illness.

Tremor

Tremor has been classified according to its etiology and to the circumstances in which the tremor occurs (Table 5). The tremor of Parkinson's disease has been discussed on page 16. Essential tremor typically affects the upper limbs, but may spread to involve the legs, head, facial muscles, voice and tongue. The tremor is sometimes asymmetrical. The condition is inherited through an autosomal-dominant gene, but also occurs sporadically. There is a bimodal age distribution with a median age of around 15 years. Alcohol relieves the tremor in approximately 50% of cases. In some patients, cogwheeling rigidity can be detected at the wrists. The tremor can readily be demonstrated by asking the patient to draw a spiral or crossed lines. Serial drawings allow an objective evaluation of drug therapy (Figure 75). The tremor sometimes responds to propranolol, phenobarbitone or primidone.

Orthostatic tremor appears on standing and affects the legs and trunk. Various tremor frequencies have been recorded in such patients, some at 6–7 Hz and others at around 16 Hz (Figure 76). Some patients display an upper-limb tremor suggestive of an essential tremor but, despite this, orthostatic tremor is more likely to respond to clonazepam than either propranolol or primidone.

Table 5 Definitions of tremor

Resting	Present when limb fully supported against gravity with the relevant muscles relaxed
Action	Present during any voluntary muscle contraction
Postural	Present during posture maintenance
Kinetic	Present during any type of movement
Intention	Exacerbation of a kinetic tremor towards the end of a goal-directed movement
Task-specific	Present during highly skilled activity such as writing or playing a musical instrument
Isometric	Present when a voluntary muscle contraction is opposed by a rigid stationary object

from Bain, 1993

Tremor is observed in a number of other situations. The tremor of cerebellar disease is typically intentional in quality, but postural elements have been described, affecting the arms at the shoulders, the legs at the hips, and the head and trunk on standing. Tremor is a recognized feature of certain neuropathies and is usually action-related. Rubral tremor is a coarse resting tremor exacerbated by posture and more so by action, and usually secondary to brain stem vascular disease or multiple sclerosis. In some dystonic syndromes, tremor appears alongside the dystonic features.

Myoclonus

This condition consists of sudden short-lived shock-like contractions of muscle. The movement varies greatly in both amplitude and frequency. Perhaps the most useful classification is anatomical, categorizing the movement as focal, segmental (two or more contiguous regions), multifocal or generalized. Although myoclonus is usually erratic in time and rhythm, it sometimes appears to be rhythmical. Some episodes of myoclonus appear spontaneously; the others appear either with startle or in response to the initiation of muscle activity.

Essential myoclonus appears in the first two decades of life and is inherited as an autosomal-dominant trait with variable penetrance. Sporadic cases are common. Postanoxic myoclonus appears after a period of coma triggered by cardiac or respiratory arrest. Muscles of the limbs, face, pharynx or trunk may be affected. Seizures are the norm, and many patients have particular problems with gait control. Drugs that enhance serotonin activity improve the condition.

Segmental myoclonus originates from a brain stem or spinal level. The movements are more or less continuous, usually at around 1–3 Hz, and explicable by discharges from contiguous anatomical levels (Figure 77). Palatal myoclonus is a rhythmic contraction of the soft palate, frequently accompanied by contraction of other muscles of the pharynx and larynx, sometimes extending to the face and even the diaphragm. Typically, it follows pontine infarction, often after a latent period of several weeks or months.

Tardive dyskinesia

Although tardive dyskinesia is typically associated with previous exposure to dopaminergic antagonists, the condition may also arise spontaneously. The movements predominate around the mouth and tongue, with lip-smacking, sucking, pursing and tongue protrusion. In some cases, involuntary movements affect the limbs or the trunk. A repetitive quality is characteristic. The condition may persist despite withdrawal of the causative agent and, indeed, may be temporarily worsened at such times. Tardive dystonia consists of focal dystonic movement particularly affecting the neck or trunk, which are also liable to persist after neuroleptic withdrawal. Both tardive dyskinesia and tardive dystonia may sometimes respond to presynaptic dopaminergic blockade with reserpine or tetrabenazine.

Selected bibliography

Anatomy

Gerfen CR, Wilson CJ. The basal ganglia. In Swanson LW, Björklund A, Hökfelt T, eds. *Handbook of Chemical Neuroanatomy, Vol. 12: Integrated Systems of the CNS, Part III.* Amsterdam: Elsevier Science BV, 1996

Riley DE, Lang AE. In Bradley WG, Daroff RB, Fenichel GM, Marsden CD, eds. *Neurology in Clinical Practice.* Boston: Butterworth–Heinemann, 1996:1734

Parkinson's disease

Hughes AJ, Daniel SE, Kilford L, Lees AJ. Accuracy of clinical diagnosis of idiopathic Parkinson's disease: A clinicopathological study of 100 cases. *J Neurol Neurosurg Psychiatr* 1992;55:181–4

Lindvall O, Sawle G, Widner H, *et al.* Evidence for long-term survival and function of dopaminergic grafts in progressive Parkinson's disease. *Ann Neurol* 1994;35:172–80

Parkinsonian syndromes

Mark MH, Sage JI, Walters AS, *et al.* Binswanger's disease presenting as levodopa-responsive parkinsonism: Clinicopathologic study of three cases. *Mov Disord* 1995;10:450–4

Stacy M, Jankovic J. Differential diagnosis of Parkinson's disease and the parkinsonism plus syndromes. *Neurol Clin* 1992;10:341–57

Gershanik OS. Drug-induced movement disorders. *Curr Opin Neurol Neurosurg* 1993;6:369–76

Fénelon G, Gray F, Wallays C, *et al.* Parkinsonism and dilatation of the perivascular spaces (*état criblé*) of the striatum: A clinical, magnetic resonance imaging, and pathological study. *Mov Disord* 1995;10:754–60

Cortical Lewy body disease

Gibb WRG, Luthert PJ. Dementia in Parkinson's disease and Lewy body disease. In Burns A, Levy R, eds. *Dementia.* London: Chapman & Hall, 1994

Gibb WRG, Esiri MM, Lees AJ. Clinical and pathological features of diffuse cortical Lewy body disease (Lewy body dementia). *Brain* 1985;110:1131–53

Mark MH, Sage JI, Dickson DW, *et al.* Levodopa-nonresponsive Lewy body parkinsonism. Clinicopathologic study of two cases. *Neurology* 1992;42:1323–7

Progressive supranuclear palsy

Perkin GD, Lees AJ, Stern GM, Kocen RS. Problems in the diagnosis of progressive supranuclear palsy. *Can J Neurol Sci* 1978;5:167–73

Daniel SE, De Bruin VMS, Lees AJ. The clinical and pathological spectrum of Steele–Richardson–Olszewski syndrome (progressive supranuclear palsy): A reappraisal. *Brain* 1995;118:759–70

Striatonigral degeneration

Gouider-Khouja N, Vidailhet M, Bonnet A-M, *et al.* 'Pure' striatonigral degeneration and Parkinson's disease: A comparative clinical study. *Mov Disord* 1995;10:288–94

Fearnley JM, Lees AJ. Striatonigral degeneration: A clinicopathological study. *Brain* 1990;113:1823–42

Multiple system atrophy

Colosimo C, Albanese A, Hughes AJ, *et al.* Some specific clinical features differentiate multiple system atrophy (striatonigral variety) from Parkinson's disease. *Arch Neurol* 1995;52:294–8

Wenning GK, Ben-Shlomo Y, Magalhâes M, *et al.* Clinicopathological study of 35 cases of multiple system atrophy. *J Neurol Neurosurg Psychiatr* 1995;58:160–6

Quinn N. Multiple system atrophy. In Marsden CD, Fahn S, eds. *Movement Disorders, Vol. 3*. London: Butterworths–Heinemann, 1994

Gilman S, Koeppe RA, Junck L, *et al.* Patterns of cerebral glucose metabolism detected with positron emission tomography differ in multiple system atrophy and olivopontocerebellar atrophy. *Ann Neurol* 1994;36:166–75

Schulz JB, Klockgether T, Petersen D, *et al.* Multiple system atrophy: Natural history, MRI morphology, and dopamine receptor imaging with [123]IBZM–SPECT. *J Neurol Neurosurg Psychiatr* 1994;57:1047–56

Corticobasal degeneration

Riley DE, Lang AE, Lewis A, *et al.* Corticobasal ganglionic degeneration. *Neurology* 1990;40:1203–12

Gibb WRG, Luthert PJ, Marsden CD. Corticobasal degeneration. *Brain* 1989;112:1171–92

Dystonia

Rothwell JC, Obeso JA. The anatomical and physiological basis of torsion dystonia. In Marsden CD, Fahn S, eds. *Movement Disorders, Vol. 2*. London: Butterworths–Heinemann, 1987

Fahn S, Marsden CD, Calne DB. Classification and investigation of dystonia. In Marsden CD, Fahn S, eds. *Movement Disorders, Vol. 2*. London: Butterworths–Heinemann, 1987

Jankovic J, Brin MF. Therapeutic uses of botulinum toxin. *N Engl J Med* 1991;324:1186–94

Wilson's disease

Scheinberg IN, Sternlieb I. *Wilson's Disease*. Philadelphia: WB Saunders, 1984

Huntington's disease

Duyao M, Ambrose C, Myers R, *et al.* Trinucleotide repeat length instability and age of onset in Huntington's disease. *Nature Genet* 1993;4:387–92

Comunale JP Jr , Heier LA, Chautorian AM. Juvenile form of Huntington's disease: MR imaging appearance. *AJR* 1995;165:414–5

Turjanski N, Weeks R, Dolan R, *et al.* Striatal D_1 and D_2 receptor binding in patients with Huntington's disease and other choreas, A PET study. *Brain* 1995;118:689–96

Sydenham's chorea

Swedo SE. Sydenham's chorea. A model for childhood autoimmune neuropsychiatric disorders. *J Am Med Assoc* 1994;272:1788–91

Tremor

Bain P. A combined clinical and neurophysiological approach to the study of patients with tremor. *J Neurol Neurosurg Psychiatr* 1993;56:839–44

Bain PG, Findley LJ, Thompson PD, *et al.* A study of hereditary essential tremor. *Brain* 1994;117: 805–24

Myoclonus

Tolosa ES, Kulisevski J. Tics and myoclonus. *Curr Opin Neurol Neurosurg* 1992;5:314–20

Deuschl G, Mischke G, Schenck E, *et al.* Symptomatic and essential rhythmic palatal myoclonus. *Brain* 1990;113:1645–72

Fahn S, Sjaastad O. Hereditary essential myoclonus in a large Norwegian family. *Mov Disord* 1991; 6:237–42

Tardive dyskinesia

Koshino Y, Madokoro S, Ito T, *et al.* A survey of tardive dyskinesia in psychiatric inpatients in Japan. *Clin Neuropharmacol* 1992;15:34–43

Gold TM, Egan MF, Kirch DG, *et al.* Tardive dyskinesia: Neuropsychological, computerised tomographic and psychiatric symptom findings. *Biol Psychiatr* 1991;30:587–99

Section 2 Parkinson's Disease and Related Disorders Illustrated

List of illustrations

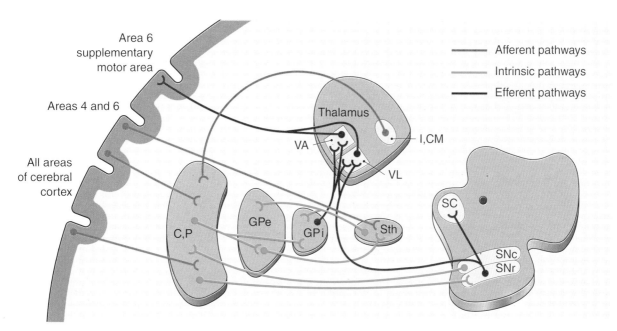

Figure 1 Major pathways of the basal ganglia (some pathways, including the subthalamonigral fibers, and afferents from the locus ceruleus and raphe nucleus, have been omitted for the sake of clarity.) C, P, caudate nucleus and putamen (striatum); GP, globus pallidus (e, externa; i, interna); SN, substantia nigra (c, compacta; r, reticulata); Sth, subthalamic nucleus; T, thalamus (nuclei: VA, ventral anterior; VL, ventro-lateral; CM, centromedian; I, other intralaminar); SC, superior colliculus. Modified from Riley and Lang, in Bradley *et al.*, *Neurology in Clinical Practice*, 1996 (see page 31)

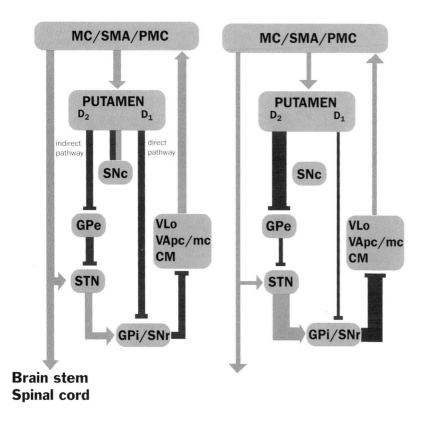

Figure 2 Connections of striatal output neurons in controls (left) and in rats with 6-0H dopamine lesions of the nigrostriatal dopamine system (right). MC, motor cortex; SMA, supplementary motor area; PMC, premotor cortex; D_1/D_2, D_1/D_2 dopamine receptor systems; SNc, substantia nigra pars compacta; SNr, substantia nigra pars reticulata; GPe/GPi, external/internal portions of globus pallidus; STN, subthalamic nucleus; VLo, ventral lateral, pars oralis, nucleus of thalamus; VApc/mc, ventral anterior, pars parvocellularis/pars magnocellularis, nucleus of thalamus; CM, centromedian nucleus of thalamus

Figure 3 Horizontal sections of midbrain (upper) and pons (lower) in idiopathic Parkinson's disease of 10 years' duration show pallor in the substantia nigra (arrowed) and locus ceruleus (arrowed), respectively

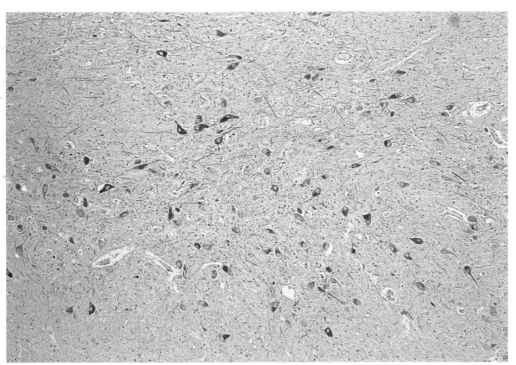

Figure 4 Histology of normal substantia nigra, which is well-populated with nerve cells immunoreactive for tyrosine hydroxylase (immunostained for tyrosine hydroxylase)

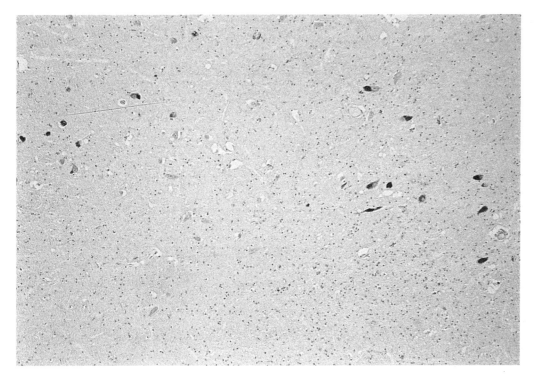

Figure 5 Histology of substantia nigra in idiopathic Parkinson's disease of 12 years' duration showing depletion of tyrosine hydroxylase-containing nerve cells (immuno-stained for tyrosine hydroxylase)

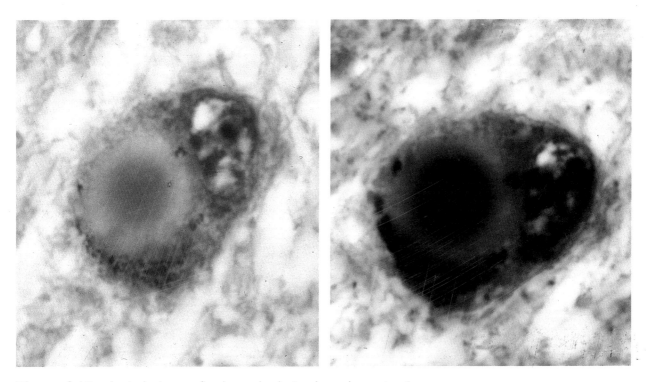

Figure 6 Histological views of a Lewy body in the substantia nigra pars compacta stained with H & E (left) and with a modified Bielschowsky stain (right)

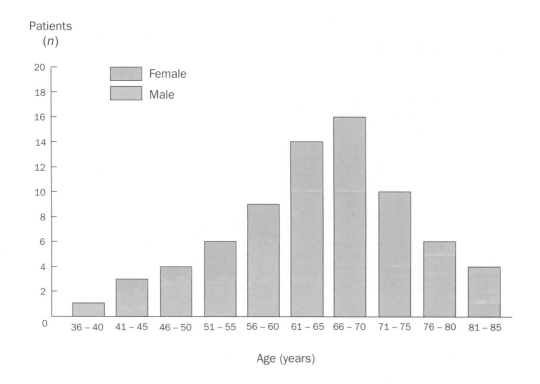

Figure 7 Age and gender distribution at the time of diagnosis in a small series of Parkinsonian patients

Figure 8 Micrographia in Parkinson's disease: The script is progressively reduced in size

Figure 9 Characteristic facial appearance in Parkinson's disease

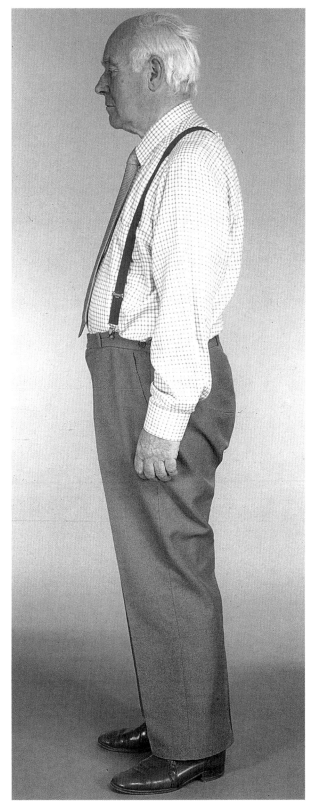

Figure 10 Posture of a patient with early Parkinson's disease

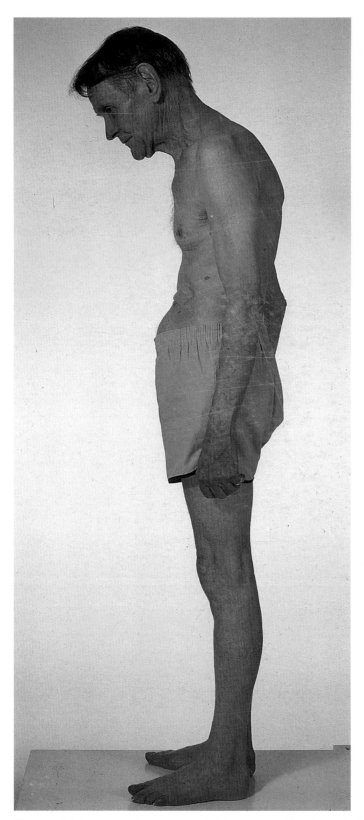

Figure 11 Posture of a patient with later-stage Parkinson's disease

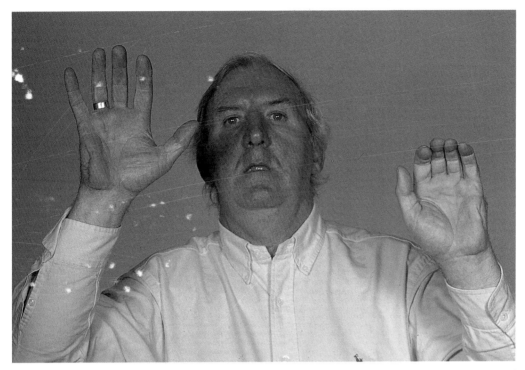

Figure 12 As this patient repetitively clenches and unclenches his fists, a paucity of movement is apparent in his left hand

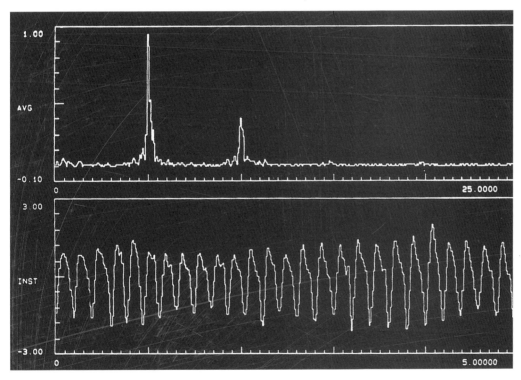

Figure 13 Power-spectrum (upper) and accelerometer (lower) tracings taken from a patient with Parkinsonian tremor. The main tremor peak is at approximately 5 Hz with a harmonic at 10 Hz

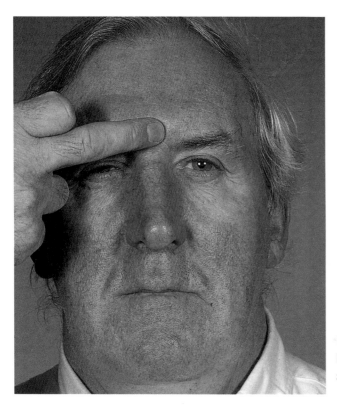

Figure 14 Positive glabellar tap. Persistent blinking is a feature of Parkinson's disease, but is also seen in Alzheimer's disease

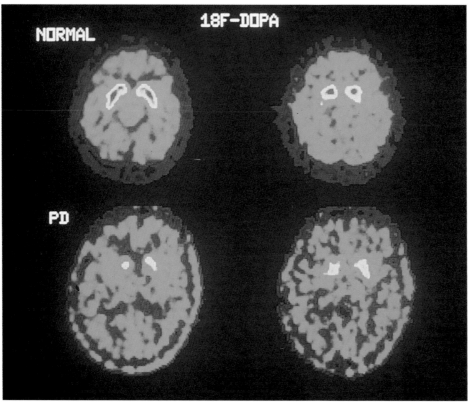

Figure 15 6-[^{18}F]-fluorodopa–PET scan appearance in a normal subject (upper) compared with a Parkinsonian patient (lower)

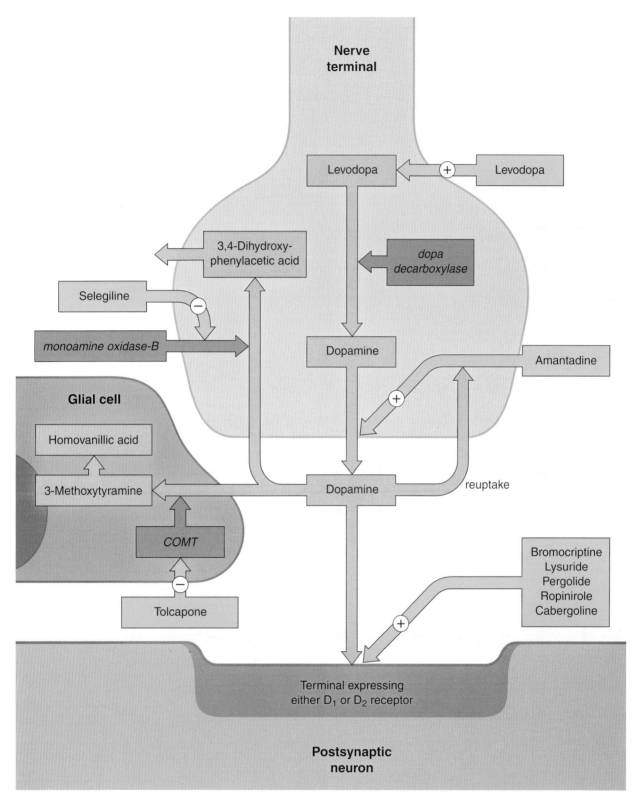

Figure 16 Synthesis and metabolism of dopamine within the central nervous system. The green arrows indicate the sites at which various agents might enhance dopaminergic activity. COMT, catechol O-methyltransferase; +, by stimulation; −, by inhibition

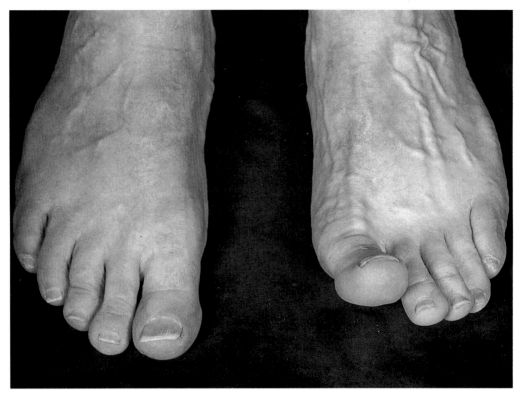

Figure 17 Dystonic posturing secondary to dopa therapy. There is hyperextension of the left big toe

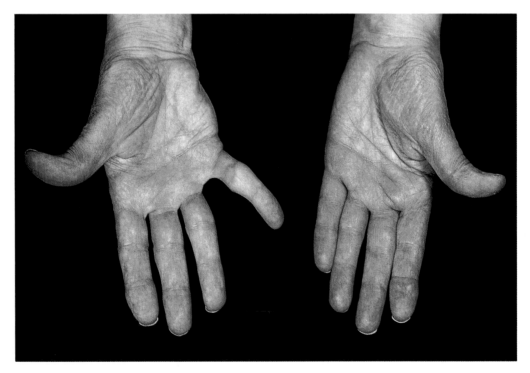

Figure 18 Dystonic posturing of the right thumb and little finger (on the left) secondary to dopa therapy

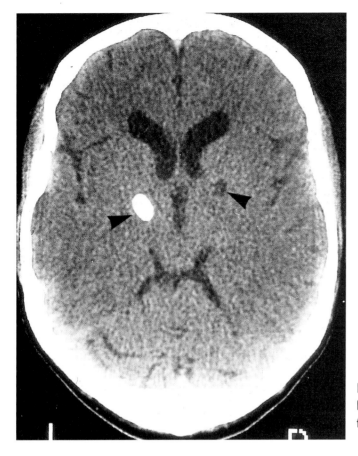

Figure 19 CT of a patient with previous bilateral thalamotomies (arrowed) performed for control of a Parkinsonian tremor

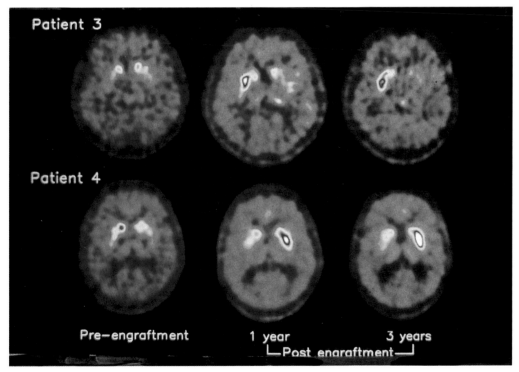

Figure 20 Fluorodopa-uptake studies in a patient following dopaminergic grafting

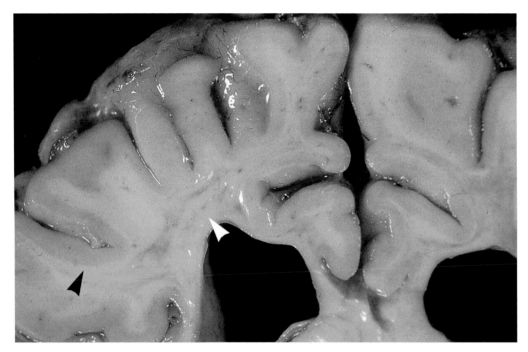

Figure 21 Coronal brain section showing abnormal white matter (white arrow) above the ventricular roof with relative preservation of subcortical white matter (U fibers; black arrow)

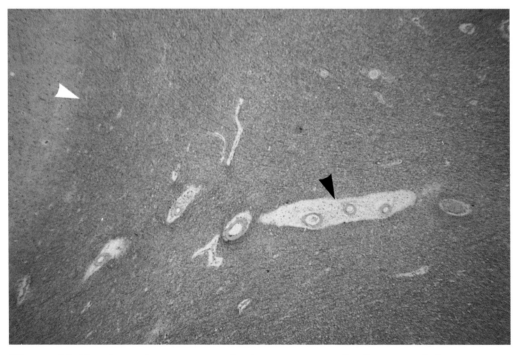

Figure 22 Histology showing parietal white matter at the bottom of the cortex, a relatively preserved (but not quite normal) arcuate zone (white arrow), and rarefied pale-staining deep white matter, containing thick-walled arteriosclerotic blood vessels lying in dilated and fibrotic perivascular spaces (black arrow)(Luxol fast blue–H & E)

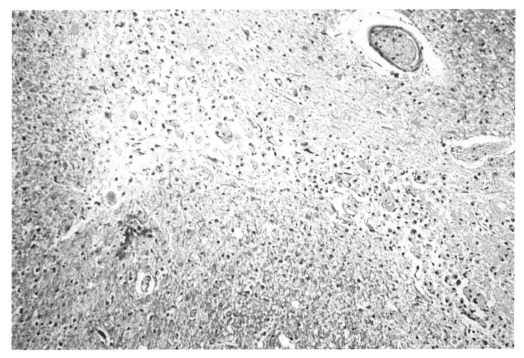

Figure 23 Histology showing a lacunar infarct (pale area) with an irregular cavity lined by reactive cells (astrocytes and macrophages). Debris and a few vascular channels can be seen (Luxol fast blue–H & E)

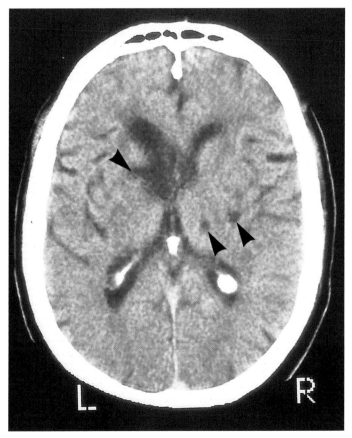

Figure 24 CT of a patient with a Parkinsonian syndrome shows multiple lacunar infarcts

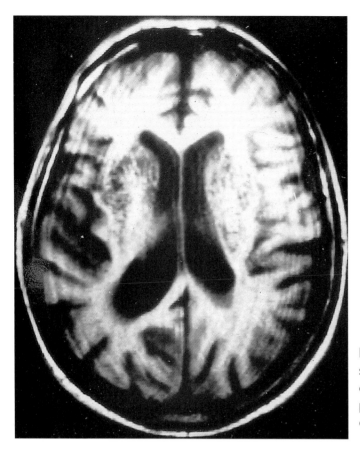

Figure 25 T₁-weighted MRI showing multiple small hypointense foci in the putamen and caudate nuclei bilaterally. The patient had presented with a Parkinsonian state with, eventually, predominant axial features

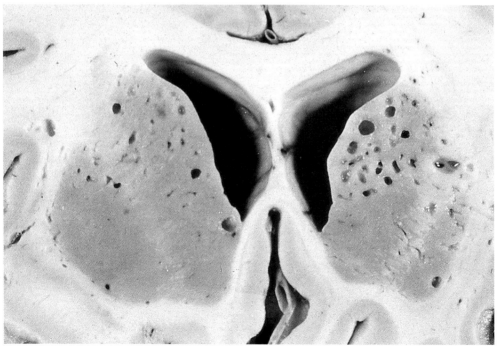

Figure 26 Coronal brain section (same patient as in Figure 25) showing numerous small lacunes in the heads of both caudate nuclei and in the anterior part of the putamen

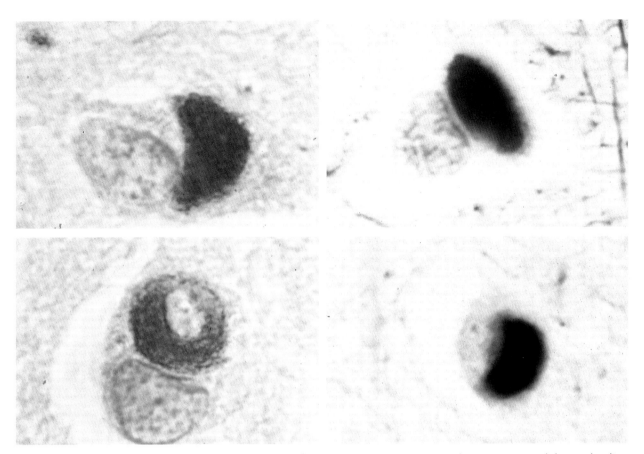

Figure 27 Histological sections from Parkinson's disease with dementia showing cortical Lewy bodies stained with ubiquitin (immunochemistry preparation, left; silver impregnation, right)

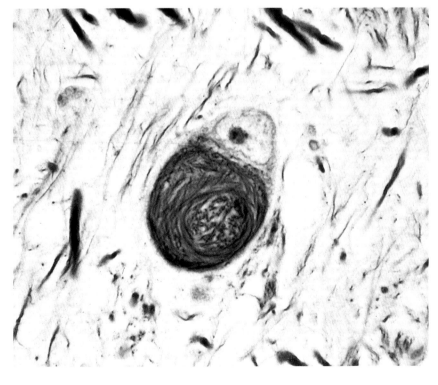

Figure 28 Histology showing a subthalamic neuron containing a globose neurofibrillary tangle in progressive supranuclear palsy (Bielschowsky silver impregnation)

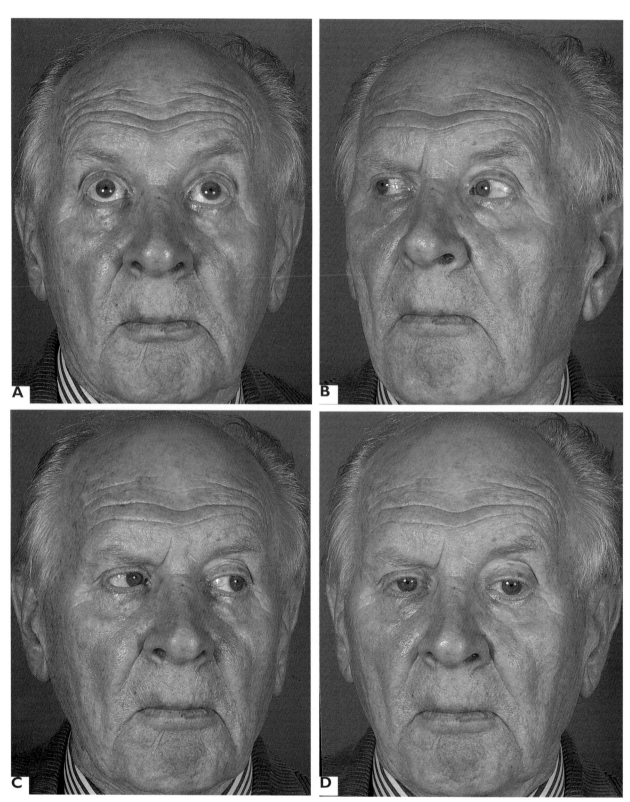

Figure 29 In this patient with progressive supranuclear palsy, upward (**A**) and lateral gaze (**B** and **C**) are preserved whereas down gaze (**D**) is impaired

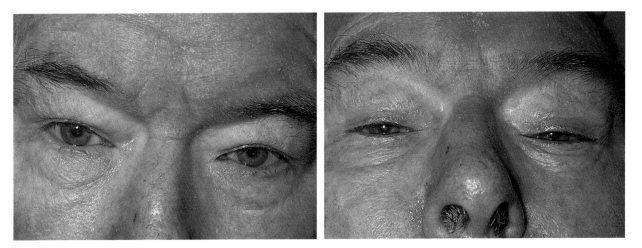

Figure 30 Attempted down gaze (left) shows improvement with the doll's-head maneuver (right) in this patient with progressive supranuclear palsy

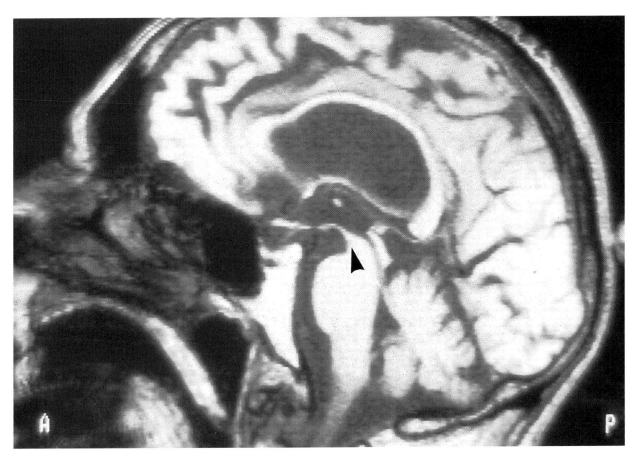

Figure 31 Sagittal T₁-weighted MRI showing midbrain atrophy (arrowed) in progressive supranuclear palsy

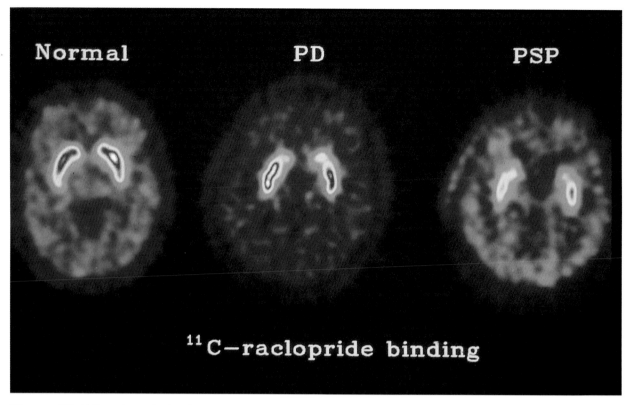

Figure 32 ^{11}C-raclopride binding in a normal subject (left) compared with that in Parkinson's disease (middle) and in progressive supranuclear palsy (right). (^{11}C-raclopride is a selective reversible antagonist of D$_2$ receptors)

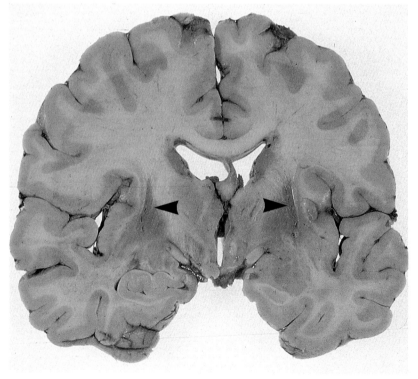

Figure 33 Coronal section of brain from a patient with striato-nigral degeneration showing symmetrical atrophy and discoloration of the putamen

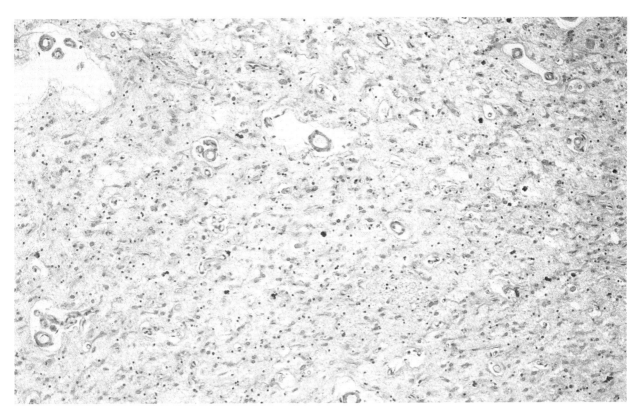

Figure 34 Histology of striatonigral degeneration shows atrophy of the putamen with rarefaction and gliosis (H & E)

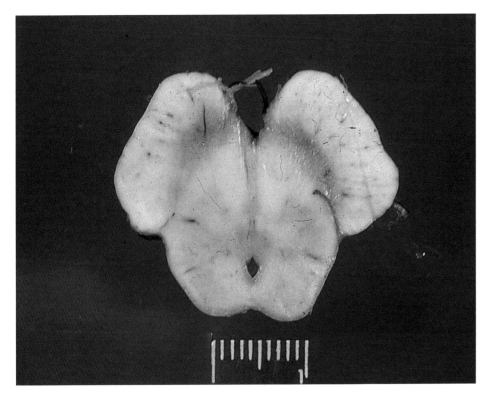

Figure 35 Transverse section of midbrain in striatonigral degeneration shows pallor of the substantia nigra

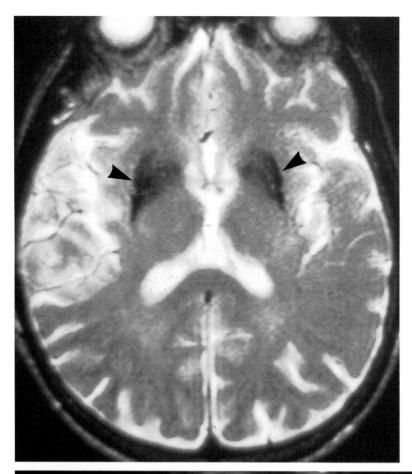

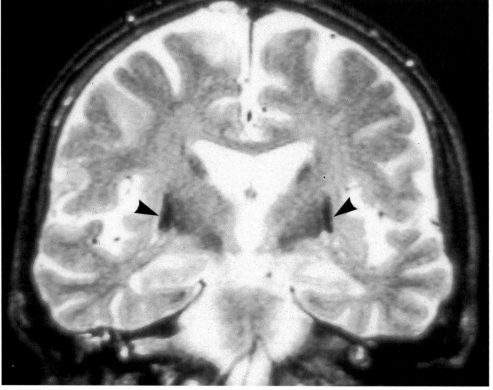

Figure 36 Axial (upper) and coronal (lower) T$_2$-weighted MRIs showing putaminal hypointensity (arrowed) in a patient with striatonigral degeneration

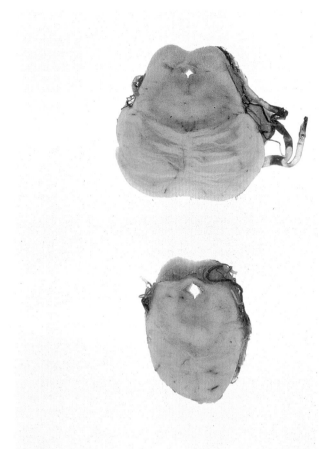

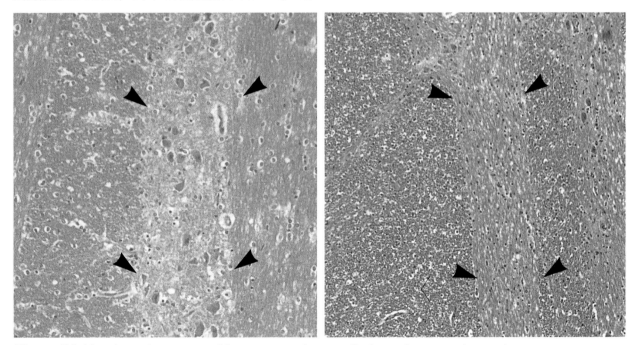

Figure 37 Coronal sections showing normal (upper) compared with atrophied (lower) basis pontis secondary to olivopontocerebellar atrophy in multiple system atrophy

Figure 38 Histological sections of basis pontis (arrowed) show the normal complement of pontine neurons in a control subject (left) compared with neuronal depletion (right) in multiple system atrophy (H & Es)

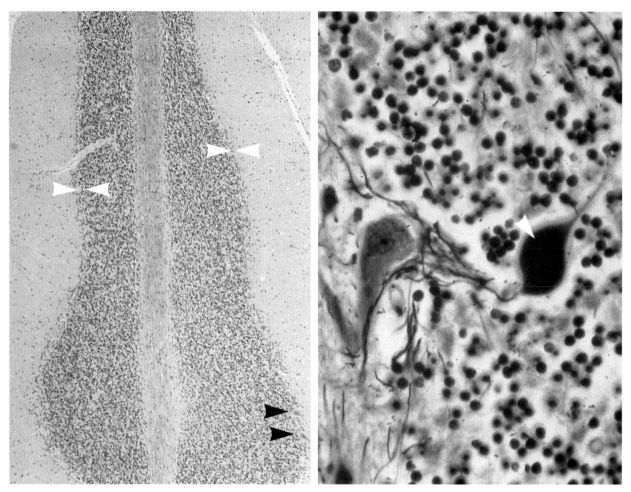

Figure 39 Multiple system atrophy with olivopontocerebellar atrophy. Histology shows (left) depletion of Purkinje cells (only two can be seen; black arrows). The remainder of the Purkinje cell layer (seen between the white arrows) consists only of small astrocytic cells. Atrophy of the central white matter in the cerebellar folia is also seen. Evidence of Purkinje cell degeneration (right) with formation of axon torpedoes (white arrow) is seen in the molecular layer (H & Es)

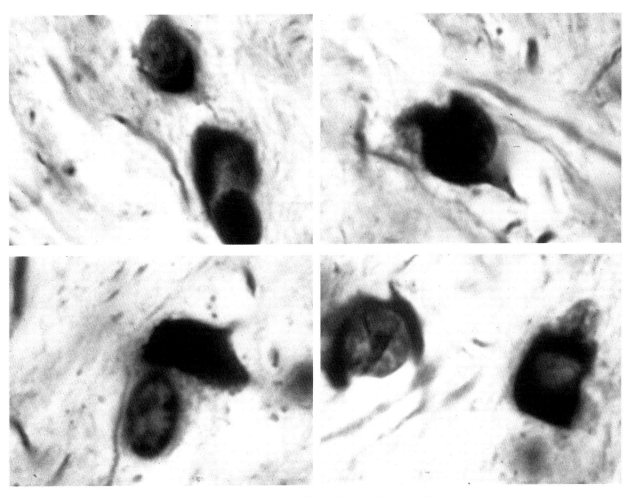

Figure 40 Histological sections showing examples of oligodendroglial cytoplasmic inclusions in multiple system atrophy (H & Es)

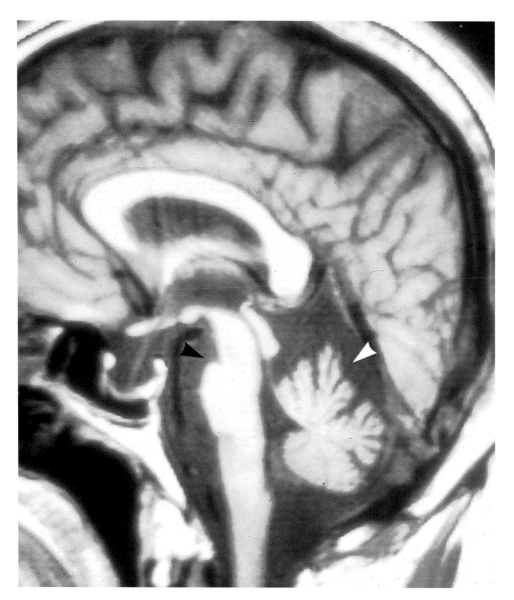

Figure 41 Sagittal T$_1$-weighted MRI showing pontine (black arrow) and cerebellar (white arrow) atrophy in a patient with olivopontocerebellar atrophy

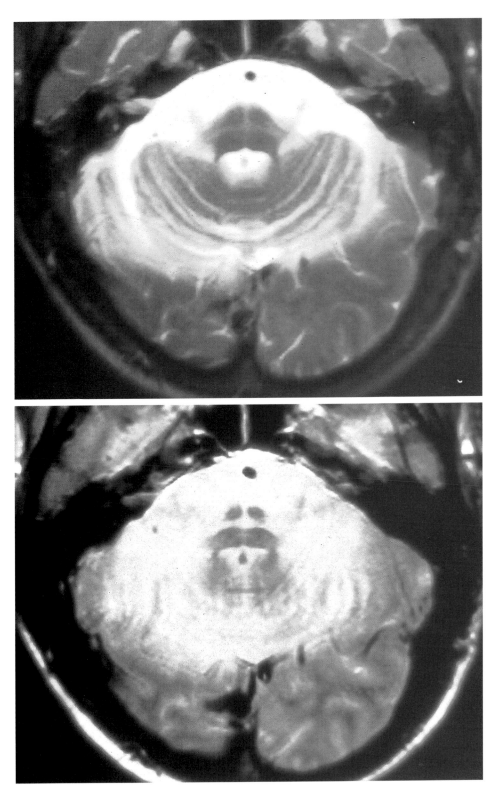

Figure 42 T$_2$-weighted MRI (upper) shows hyperintensity of the middle cerebellar peduncles and the cerebellum. The axial proton-density MRI (lower) shows hyperintensity of the transverse pontine fibers, middle cerebellar peduncles and cerebellum in olivopontocerebellar atrophy

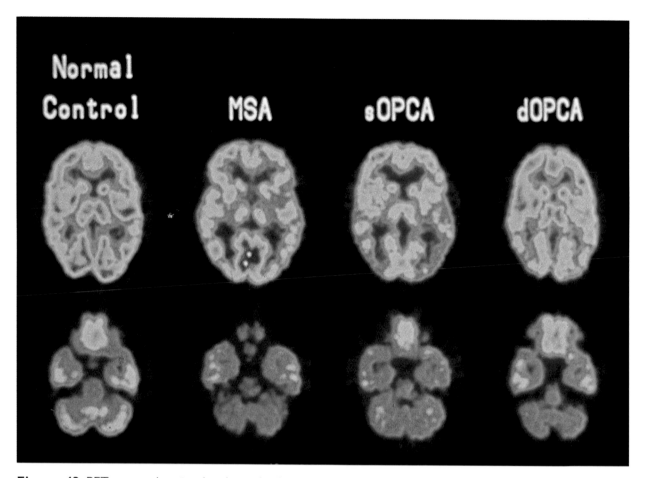

Figure 43 PET scans showing local cerebral metabolic rate for glucose in a normal control compared with patients with multiple system atrophy (MSA), sporadic olivopontocerebellar atrophy (sOPCA) and dominantly-inherited olivopontocerebellar atrophy (dOPCA)

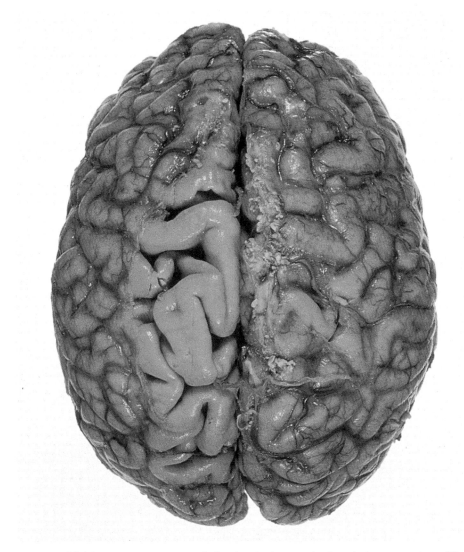

Figure 44 Macroscopic view of the brain in corticobasal degeneration. The leptomeninges have been partially removed to show cortical atrophy involving the posterior frontal and anterior parietal regions

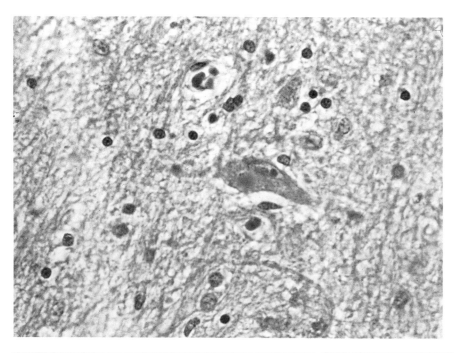

Figure 45 Histological section of cerebral cortex in corticobasal degeneration shows severe gliosis and an achromatic neuron (H & E)

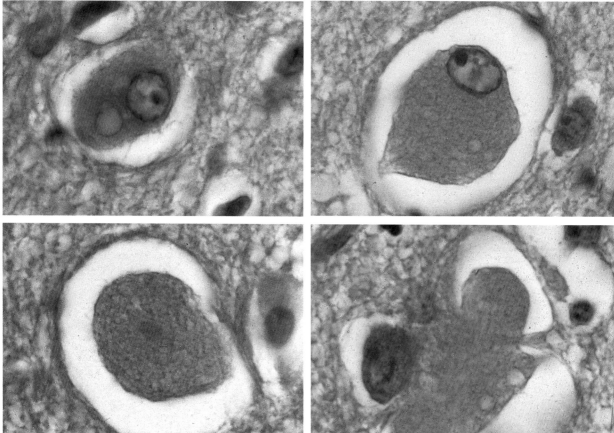

Figure 46 Histological sections of cerebral cortex showing swollen cortical neurons in corticobasal degeneration with an appearance resembling chromatolysis (H & Es)

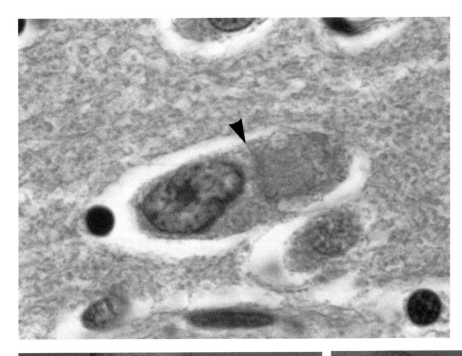

Figure 47 Histology of cerebral cortex in corticobasal degeneration shows a putaminal neuron (arrowed) containing a basophilic inclusion (H & E)

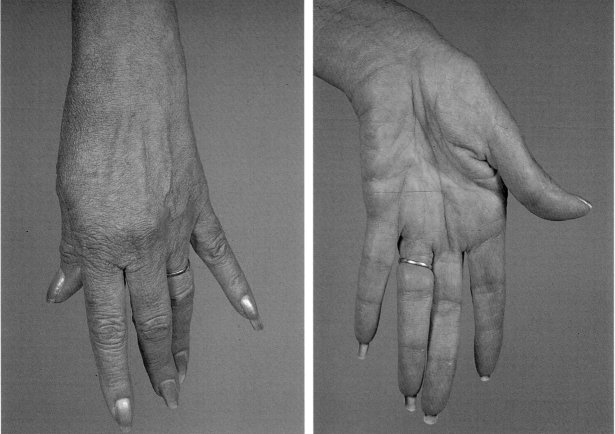

Figure 48 Dorsal (left) and palmar (right) views of dystonic posturing of the left hand of a patient with corticobasal degeneration. In particular, note the ulnar deviation at the wrist and the abducted posture of the little finger

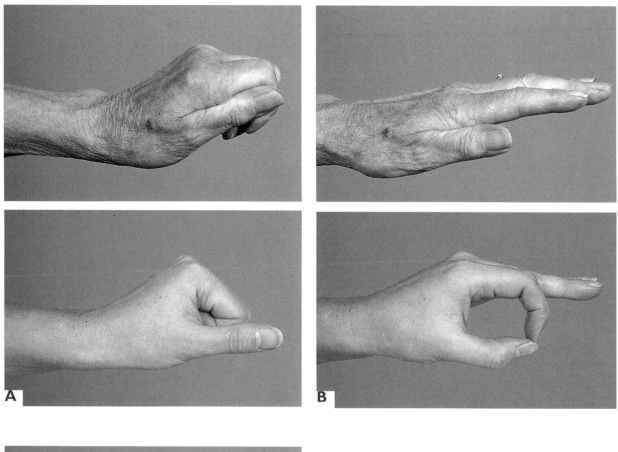

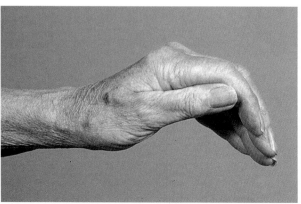

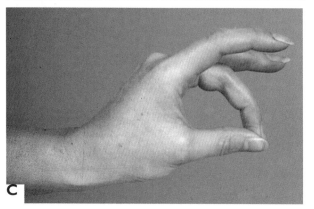

Figure 49 This patient with corticobasal degeneration shows ideomotor apraxia of the left hand. When asked to copy three hand postures (**A–C**, lower), in each instance, the patient's version was defective (**A–C**, upper)

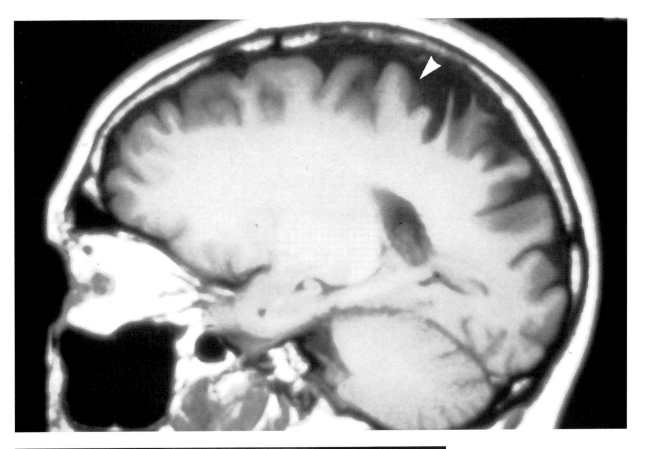

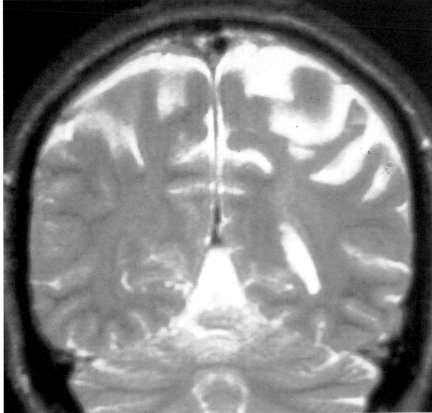

Figure 50 In this patient with corticobasal degeneration, sagittal T_1-weighted MRI (upper) shows predominantly posterior frontal and parietal atrophy (arrowed). Coronal T_2-weighted MRI (lower) shows that the parietal atrophy is asymmetrical

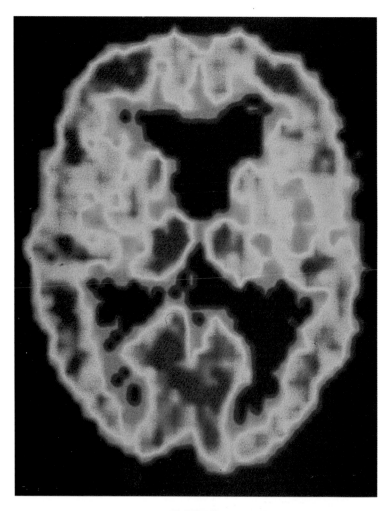

Figure 51 [^{18}F]-fluorodeoxyglucose–PET scan shows reduced metabolism in the left frontoparietal cortex and left striatum in a patient with corticobasal degeneration

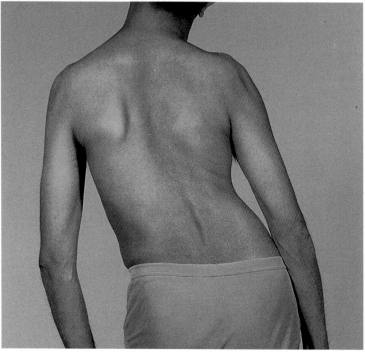

Figure 52 Scoliotic posture in a patient with idiopathic torsion dystonia

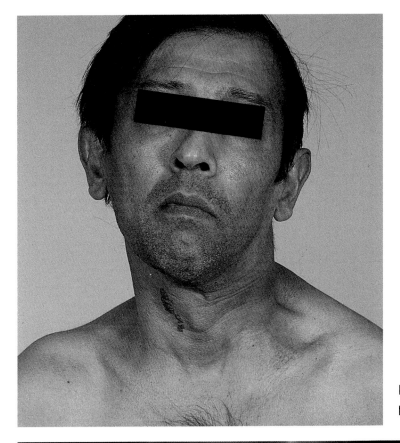

Figure 53 Abnormal neck posture in a patient with idiopathic torsion dystonia

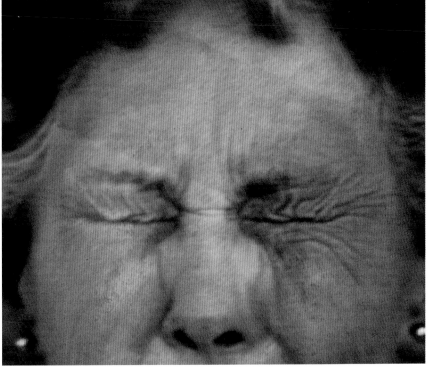

Figure 54 Blepharospasm: Still photograph taken from a videorecording. The condition proved responsive to botulinum toxin

Figure 55 Abnormal neck posture in a patient with spasmodic torticollis

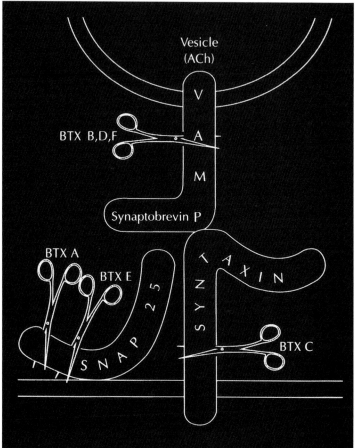

Figure 56 Mechanisms of actions of various botulinum toxins (BTX). Ach, acetylcholine

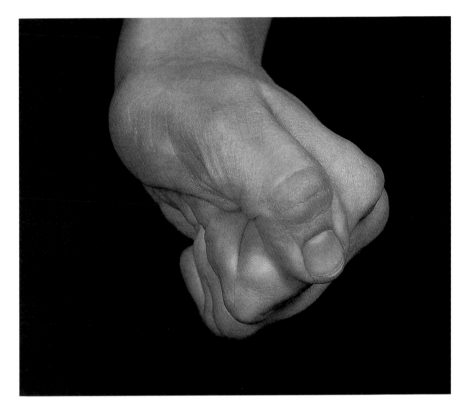

Figure 57 Dystonic posturing of the hand consequent to perinatal hypoxia

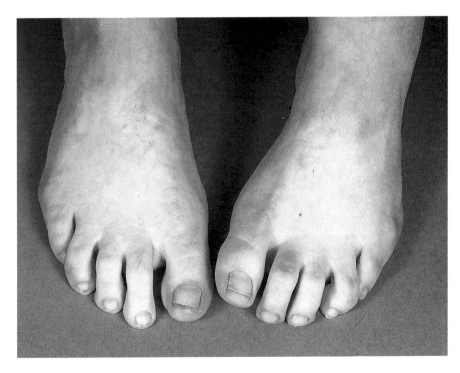

Figure 58 Dystonic posturing of the foot consequent to perinatal hypoxia. There is inversion of the foot and relative dorsiflexion of the big toe of the patient's left foot compared with the right

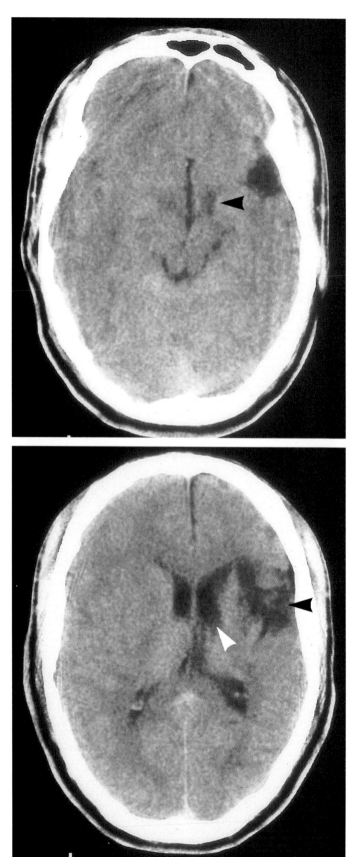

Figure 59 CTs showing focal ischemic change (black arrows) in the right frontal region with deep extension and dilatation of the right frontal horn (white arrow)

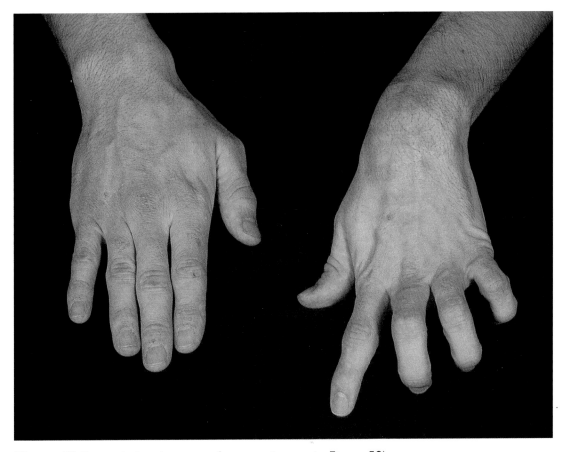

Figure 60 Dystonic hand posture (same patient as in Figure 59)

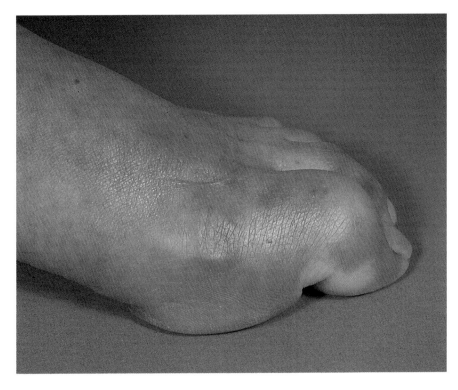

Figure 61 Dystonic posturing of the left big toe in a patient who had experienced an ischemic event in the right cerebral hemisphere following attempted removal of a sphenoidal wing meningioma

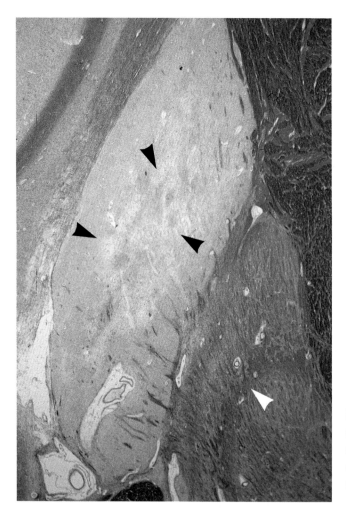

Figure 62 Histology in Wilson's disease shows atrophy and rarefaction of the putamen (black arrows). Loss of myelin bundles in the putamen contrasts with the more normal appearances in the globus pallidus (white arrow; Luxol fast blue)

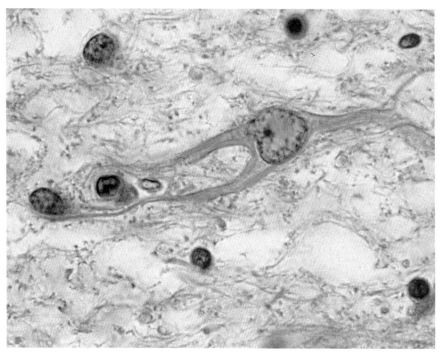

Figure 63 Histology of brain in Wilson's disease shows a Bergmann type 2 astrocyte within an atrophic putamen (H & E)

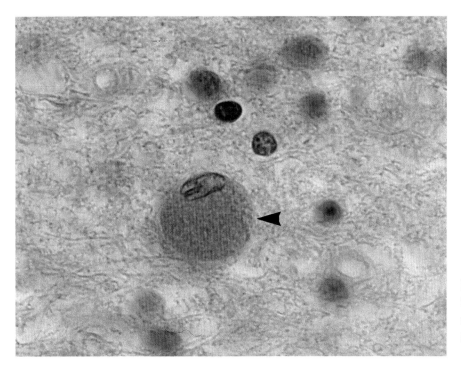

Figure 64 Histology of brain in Wilson's disease showing an Opalski cell (arrowed) (H & E)

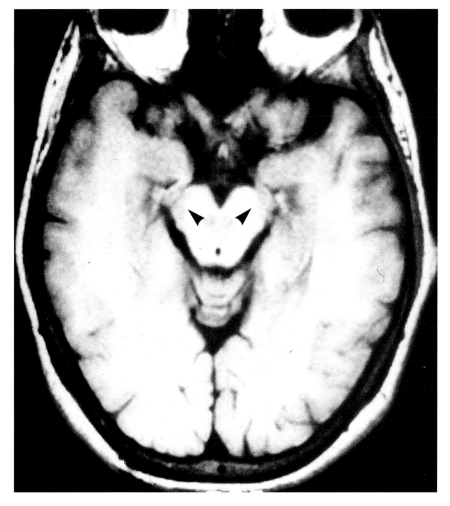

Figure 65 T$_1$-weighted MRI shows the presence of high-signal areas (arrowed) in the substantia nigra in a patient with chronic acquired hepato-cerebral degeneration

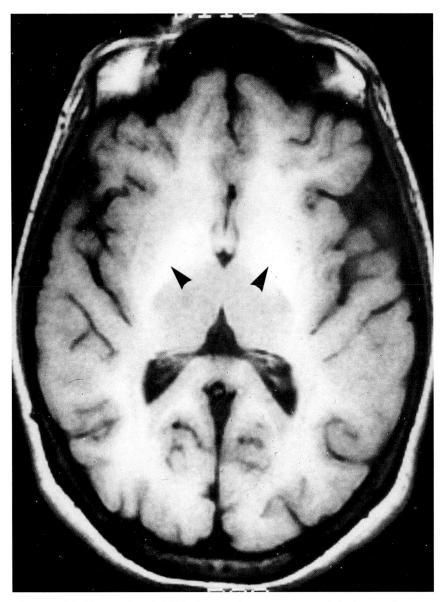

Figure 66 T$_1$-weighted MRI shows the presence of high-signal areas (arrowed) in the pallidum in a patient with chronic acquired hepatocerebral degeneration

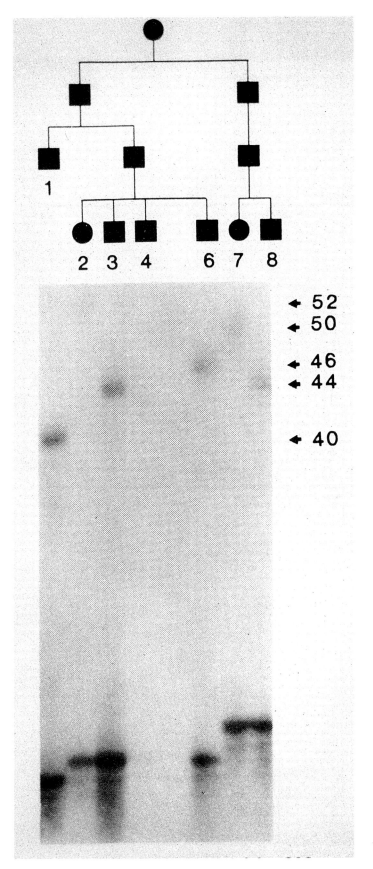

Figure 67 Pedigree (upper) of a family with Huntington's disease is accompanied by DNA gels (lower) which indicate trinucleotide repeats of between 40 and 52

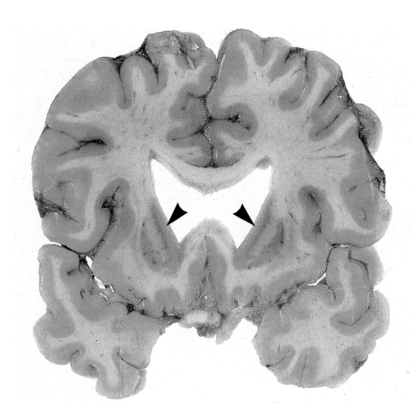

Figure 68 Coronal section of brain in Huntington's disease shows symmetrical atrophy and brown discoloration (arrowed) of the caudate and putamen together with dilatation of the lateral ventricles

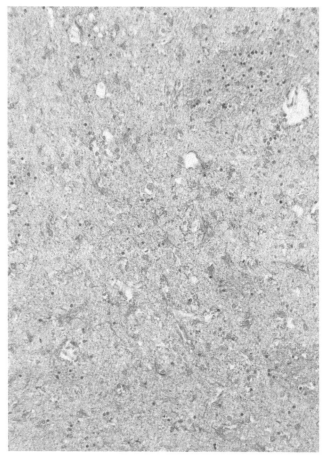

Figure 69 Histological section of brain in Huntington's disease shows atrophy with loss of neurons and astrocytic gliosis (immunocytochemistry preparation for glial fibrillary acidic protein)

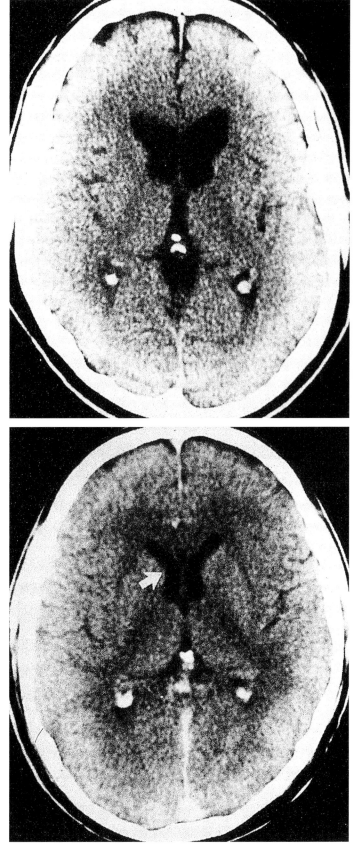

Figure 70 CTs showing atrophy of the caudate nucleus in a patient with Huntington's disease (upper) compared with a normal subject (lower; arrowed)

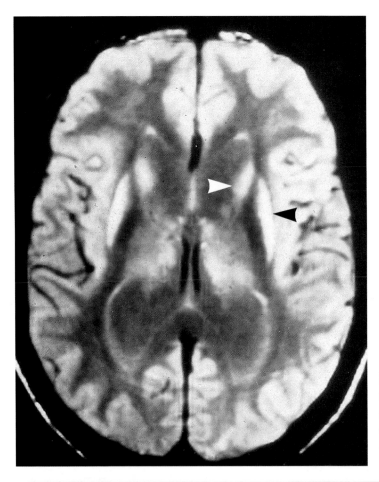

Figure 71 Axial proton-density MRI in a patient with Huntington's disease shows areas of increased signal in both the caudate nucleus (white arrow) and putamen (black arrow)

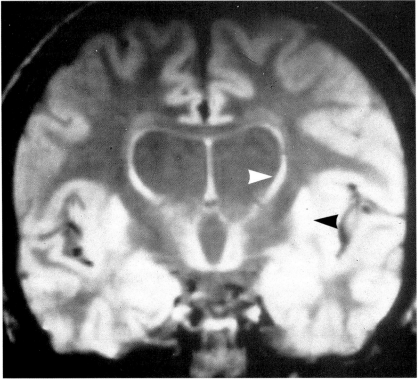

Figure 72 Coronal proton-density MRI in a patient with Huntington's disease shows features similar to those in Figure 71

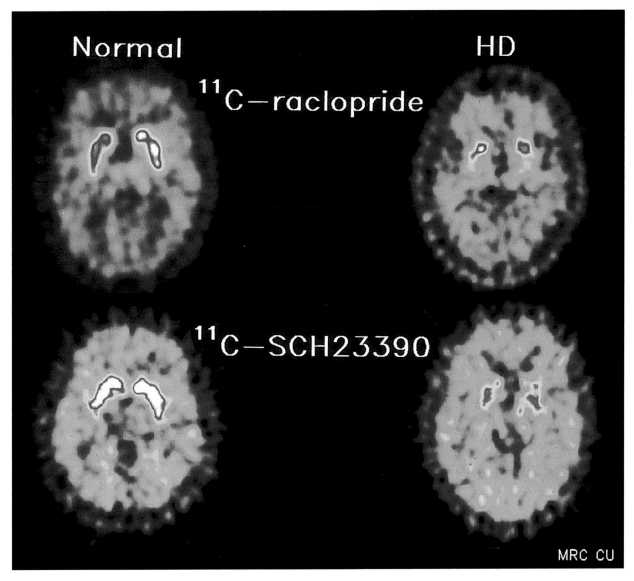

Figure 73 PET scans showing integrated ¹¹C-raclopride and ¹¹C-SCH 23390 activity in a normal subject (left) and choreic patient (right) with Huntington's disease. Both D_1 and D_2 binding is reduced in the Huntington's patient in both the caudate and putamen

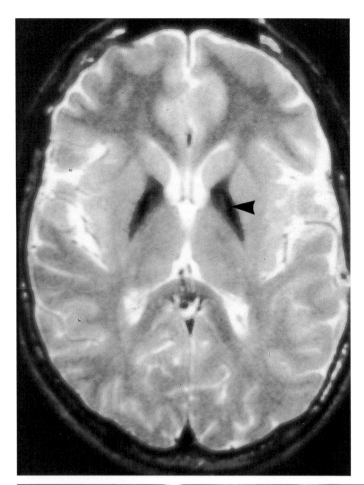

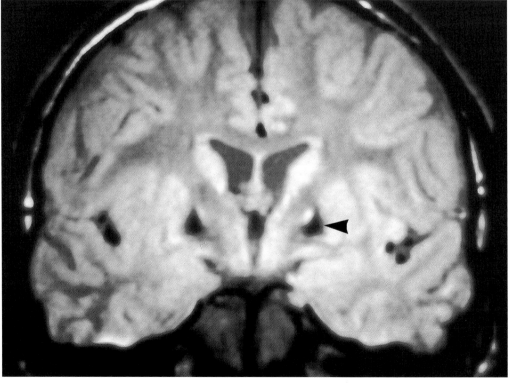

Figure 74 Axial T$_2$-weighted MRI (upper) of a patient with Hallervorden–Spatz disease shows marked pallidal hypointensity (arrowed). The coronal proton-density MRI (lower) shows a similar picture together with an anteromedial zone of high intensity

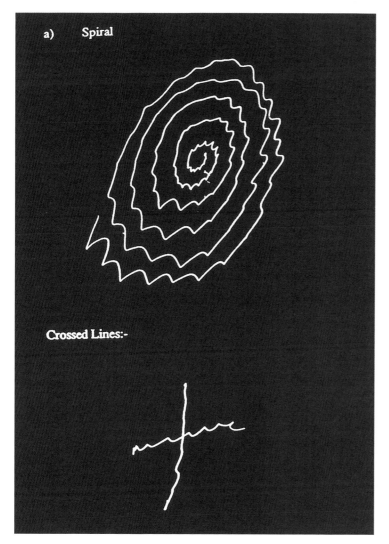

a) Spiral

Crossed Lines:-

Figure 75 An Archimedean spiral and cross drawn by a patient with an essential tremor

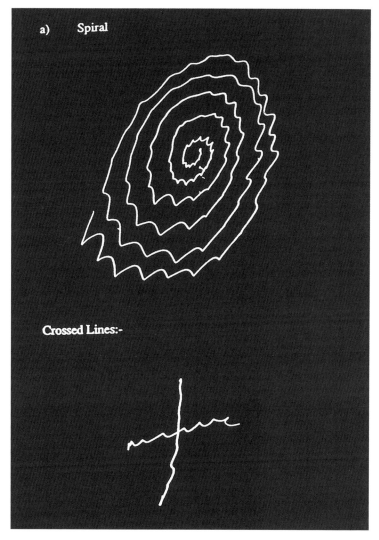

Figure 75 An Archimedean spiral and cross drawn by a patient with an essential tremor

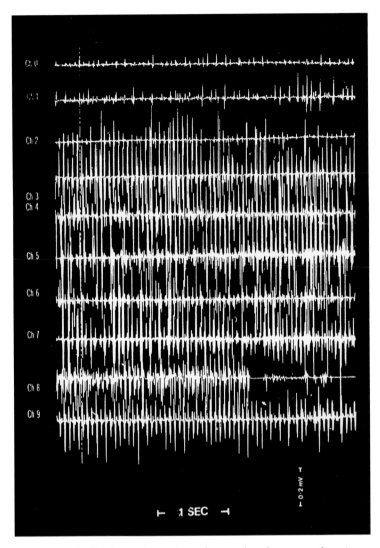

Figure 76 Surface recording from the leg muscles in a patient with primary orthostatic tremor shows a tremor of approximately 16 Hz

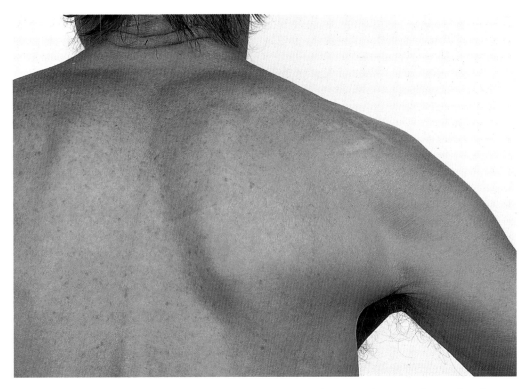

Figure 77 Irregular contractions of the periscapular muscles on the right due to spinal myoclonus

Index

The page numbers in **bold** refer to illustrations.